ıdio

Technician Class Licensing

for 2018 through 2022

License Examinations

Stephen Horan, NM4SH

US Amateur Radio Licensing Series

Amateur Radio Technician Class Licensing
for 2018 through 2022 License Examinations

First printing: April 2018

ISBN-13: 978-1986828611

Cover image by the author.

Author

Stephen Horan, NM4SH, has been a licensed amateur for over 20 years. During that time, he helped develop and lead weekend licensing classes with the Mesilla Valley Radio Club in Las Cruces, NM where he was also a Volunteer Examiner. He mostly operates on digital modes from his home in Virginia where he is member #1501 of the PODXS 070 Club and has Loyal Order of Narrow-banded Phase-shifters (LONP) certificate #266.

Professionally, he has over 20 years' experience as a professor of Electrical Engineering at New Mexico State University. He presently works as an engineer for NASA. In addition to this series of Amateur Radio licensing study guides, he is the author of the textbook *Introduction to PCM Telemetering Systems* and has published numerous technical articles in journals and conferences.

Visit his author page is at `https://amazon.com/author/shoran`. You can reach him via e-mail at `nm4sh@arrl.net`.

Also by the Author in this Amateur Radio Licensing Series:

- Amateur Radio Extra Class Licensing
- Amateur Radio General Class Licensing

Contents

List of Figures

List of Tables

BACKGROUND

Note to Readers

Welcome to the world of Amateur Radio. Literally, this is a welcome to the world because, with amateur radio, you will have access to a community of friends from around the world. But before you can fire up your equipment, you must first hold a license to share the radio bands with both other amateurs and other users in that world community. Each country has its own licensing process and I have designed this study guide to help you through the process used in the United States (US). The amateur community has standardized the licensing questions so everyone seeing the amateur radio license test in the US will see questions drawn from the same master pool. Amateurs participating in the National Conference of Volunteer Examiner Coordinators (NCVEC) `http://www.ncvec.org` design and maintain the question pool. This version of the question pool is valid from 1 July 2018 through 30 June 2022. The administrators for your license examination will be other amateur radio operators who have certifications to give the examinations. They do this as volunteers. They have all been through this same process, so you will be among friends.

It is essential that you keep in mind the purpose of the study guide: to enable prospective amateurs, with no prior knowledge of electronics and amateur radio, to pass the entry-level Technician examination. This means a certain level of compromise, but you will not need an engineering degree to be successful. For this study guide format, I assume that you are coming with your natural enthusiasm, but without a large amount of technical preparation beforehand. As much as I would like to imbue you with a full and comprehensive understanding of every technical, operational, and regulatory element of amateur radio, we must accept the limitation of a study-guide based approach and concentrate on the knowledge you will need to pass the exam. Each chapter covers one part of the Technician question pool. The chapter starts with a short discussion of key concepts that are behind several of the questions in the chapter that need more context than the question responses provide. Then, we proceed into the questions and answers with the correct answer identified and explained why it is correct. In many cases, the explanations for the answers go beyond factually picking the right option among the answers given to include why the incorrect answers are wrong so that you will learn as we go through the questions. I have tried out many of the explanations for the right and wrong responses on prospective amateurs during weekend-long preparation sessions. I am striving to provide sufficient instruction through the process to ensure successful completion of the Technician Class exam element and learn about amateur radio practice. Once you have successfully passed the examination and have your "ticket," I encourage you to pursue the theory and practice of amateur radio more deeply and work towards more advanced licenses that give you greater operating privileges.

"How to Study" Suggestions

Given that you are doing this preparation amid all your other activities, I offer the following suggestions for being more efficient in the use of your time.

Become familiar with the material Before you can study effectively, you need to know where you are going! To do this, look over the question pool and the explanations. The general format is the same on all amateur radio exams. See that there are "technical parts," "operational parts," "safety parts," and "rules and regulations parts." Get a general sense for the flow of the material and the level of detail required. For the Technician exam, the level of electronics and mathematics is relatively simple, so you should have no real difficulties in passing these topics.

Assess what you already know and what is new Based on your experience and knowledge (if any) of amateur radio, parts of the material will look familiar. Other parts, perhaps the detailed technical concepts, will be new. For now, try to concentrate on the new concepts and put less emphasis on the familiar concepts. Read through the familiar concepts to understand what you need to learn, but do not put most of your energy there. Try to identify the areas that represent new concepts to you and place most of your energy there.

Prioritize what is new to you Depending upon your background and interests, you will easily grasp some of the new concepts and others will leave you really wondering what it's all about. Sort out the topic areas by level of difficulty to you. Give yourself some confidence by working through the concepts that seem easier to you and build yourself towards the more difficult concepts. Use study aids to help you track the key concepts that you need to learn and help you to become familiar with the concepts.

Make a list of questions For those concepts that are really causing you difficulty, write them down. Consult resources you have access to such as friends who are amateurs, Web references in the study guide, or even Wikipedia. If there is an amateur radio club in your area, there will be people there ready to help you.

Make study aids You can learn some items, for example the frequency allocations, either through experience or by memorization. Since you do not have any operating privileges at this time, it is hard to have the experience, so you will most likely need to memorize them. To make the memorization easier, make yourself a set of index cards with the allowable frequencies, power limits, and types of emissions allowed. You can also do this with the necessary equations for antenna design, power densities, or other concepts that you are trying to master. Review the index cards a few times each day until you feel more comfortable with the concepts. Do not try to memorize everything all in one sitting.

Test yourself There are many on-line practice exams you can take to see if you are ready for the real thing. Visit sites such as `https://www.qrz.com/hamtest/`, `http://www.arrl.org/exam-practice`, or `http://www.eham.net/exams/`. You can find more sites by using a search engine in your Web browser. When you score consistently above 75 % you should be ready.

Relax Remember, you do not need to get a 100% to pass the exam! Do as well as you can in learning the concepts that you can grasp. I have designed the commentary on the questions to help you understand which of the four choices for an answer is correct and which is incorrect with a bit of explanation why. Learn to recognize the distractor answers when reviewing the question pool. I have phrased some of the comments to help you eliminate these distractor choices. Do not put yourself under pressure to memorize everything. Be willing to tell yourself that you can write off a few questions that you just cannot get at this time and hope for the best when you get the real exam. You will have plenty of time after you get your Technician ticket to master the difficult concepts once you become "radioactive."

About the Exam

The Federal Communications Commission (FCC) mandates that the question pool for the Amateur Radio Service license examinations have a certain structure. The NCVEC has a question pool committee that designs and publishes the questions used. The general method is to break the overall question pool into three major license elements with each element covering radio theory, operations, safety, and regulations. The three major elements are:
Element 2 — Technician Class License
Element 3 — General Class License
Element 4 — Extra Class License
Note: there no longer is an Element 1; that was the Morse code examination.

The question pool designers then break the license elements into ten subelements. They next divide the subelements into a varying number of question groups, depending upon the question pool designers' curriculum. The Technician question pool has over 400 questions divided into ten subelements. Each subelement will have between 2 and 6 groups of at least 10 questions. Each license exam for Elements

2, 3, and 4 will use one and only one question from each group regardless of the number of questions in the question pool for that group. Use this knowledge to help you to design your studying. Obviously, you should not try to memorize each question, but to learn the general principles.

Table 1 shows how the designers organized the question pool for the Technician Class license Element 2. In the individual chapters, where we look at the questions, you will see a code at the start of each question. The Technician exam will have a total of 35 questions with one question drawn randomly from each of the groups. Each question has a code that looks like **T1A01**. You can decode this as

T1 — Element 2 (Technician), subelement 1
A — Group A of subelement 1
01 — Question 1 from Group A

The questions appear in this study guide in the same format as presented by the question pool committee. On the actual exam that you take, the wording of the questions and the answers will be the same, but the examination designer will randomize the order of question groups and each question's answers. Be sure to read each question and the four possible answers carefully before selecting the right answer.

Exam Day

A Volunteer Examiner team administers license exams at pre-scheduled, publicly announced test sessions. Oftentimes, you will need to pre-register with the team before taking the exam. Be sure to understand the date, time, location, and registration requirements for the session.

The exam will only cover the Element 2 questions from the Technician Class license pool. If you are feeling especially motivated, you may then take the General Class exam, Element 3, after you have successfully completed the requirements for the Technician Class license. If you pass both these, you can even take the Amateur Extra exam as well at this session!

What do you need to bring with you for the exam? Be sure to have the following items physically with you when the exam starts:

- several sharpened pencils and an eraser
- a photo-ID or other valid forms of identification (ask the examination team what alternatives they accept if a photo-ID, such as a driver's license, is not available)
- your social security number or, if you have it, your FCC-issued Federal Registration Number
- you may also wish to bring a calculator; be sure that you have not stored exam-related formulas or data in the calculator's memory because the examination team will check to see that it is empty

Verify this list with the examination team either when you register or before the testing session to make sure the session has no other restrictions.

You will have all the time you need to complete the exam. Do not rush. Read each

Table 1: 2018 – 2022 Technician Question Pool Organization

Subelement	Content	Groups	Questions
T1	FCC Rules, descriptions, and definitions for the Amateur Radio Service; operator and station license responsibilities	6	6
T2	Station operation: choosing an operating frequency; calling another station; test transmissions; procedural signs; use of minimum power; choosing an operating frequency; band plans; calling frequencies; repeater offsets	3	3
T3	Radio wave characteristics: properties of radio waves; propagation modes	3	3
T4	Amateur radio practices and station set up	2	2
T5	Electrical principles: math for electronics; electronic principles; Ohm's Law	4	4
T6	Electrical components; circuit diagrams; component functions	4	4
T7	Station equipment: common transmitter and receiver problems; antenna measurements; troubleshooting; basic repair and testing	4	4
T8	Modulation modes: amateur satellite operation; operating activities; non-voice and digital communications	4	4
T9	Antennas and feed lines	2	2
T0	Electrical safety: AC and DC power circuits; antenna installation; RF hazards	3	3
		35	35

question carefully and be sure to indicate the correct answer. There is no penalty for guessing. If you must guess, try to eliminate as many choices as possible for the question and then select the remaining option that seems the most correct to you.

Chapter 1

T1 – RADIO RULES AND REGULATIONS

1.1 Introduction

We begin our study with the rules and regulations that govern the Amateur Radio Service. We will concentrate mostly on factors affecting amateurs in the United States (US), but we will also see some international rules as well. One mistake that many license candidates make, especially those with a natural science or engineering background, is to try and understand where some of these regulations come from like they would a physical law. That is futile! It is best to understand the general flow and agree that you will need to memorize some of these rules.

The Federal Communications Commission (FCC) (`https://www.fcc.gov`) is the federal entity that governs the Amateur Radio Service in the US. The Code of Federal Regulations (CFR), in the section called Title 47 — Telecommunication, captures the FCC rules. Title 47 has five volumes comprising four chapters and you can find it on the Web at `https://www.ecfr.gov/cgi-bin/text-idx?tpl=/ecfrbrowse/Title47/47tab_02.tplthe`. The Amateur Radio Service is one of many radio services in Title 47 and Part 97 covers it. You can find the full text of Part 97 by following the link in Title 47. The Amateur Radio Service is located within the Wireless Telecommunications Bureau at the FCC.

The FCC does not operate independently in the world. Radio communications do not stay within national borders, so the rules and regulations must be compatible with telecommunications use by other nations. National equivalents of the FCC work with their peer government agencies through the International Telecommunication Union (ITU) to bring some degree of order to the interests. The ITU divides the Earth into three regions as Figure 1.1 shows. The states in the US are in Region 2. We have special license arrangements within Region 2 with our immediate neighbors, as you might expect. The URL for the ITU is `http://www.itu.int/en/Pages/default.aspx`. A final Web site you may found useful is for the International Amateur Radio

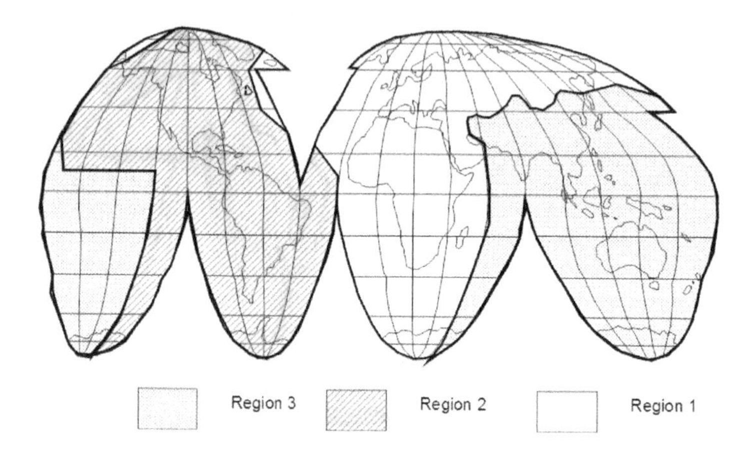

Figure 1.1: The ITU regions around the world. The U.S. is in ITU Region 2.

Union (IARU), which you can find at the URL `http://www.iaru.org`.

The questions in Subelement 1 will select from topics throughout Part 97. These cover some of the most basic concepts dealing with radio communications and the associated regulations. As we progress in our license studies to the General and Extra Class, we will get more into the details of Part 97. Do not try to memorize all of Part 97, but read through it to understand the general flow of concepts. As we look at the questions and answers in this chapter, we will see quotes from Part 97 that are relevant. Next to the questions, the question pool also lists the specific section in Part 97 where the question originates. This Part 97 information will not appear on the exam questions – it is only for study purposes.

This *Radio Rules and Regulations* subelement has the following question groups:

A. Amateur radio service
B. Authorized frequencies
C. Operator licensing
D. Authorized and permitted transmission
E. Control operator and control types
F. Station operations

This will generate six questions on the Technician examination.

1.2 T1A – Amateur Radio Service

1.2.1 Overview

The *Amateur Radio Service* question group in Subelement T1 covers the definition of the Amateur Radio Service in the FCC rules. The *Amateur Radio Service* questions cover

- Purpose and permissible use of the Amateur Radio Service

- Operator/primary station license grant
- Meanings of basic terms used in FCC rules
- Interference
- Radio Amateur Civil Emergency Service (RACES) rules
- Phonetics
- Frequency coordinators

There is a total of 11 questions in this group of which one will be selected for the exam.

1.2.2 Questions

T1A01 [97.1] Which of the following is a purpose of the Amateur Radio Service as stated in the FCC rules and regulations?

A. Providing personal radio communications for as many citizens as possible
B. Providing communications for international non-profit organizations
C. Advancing skills in the technical and communication phases of the radio art
D. All of these choices are correct

When we inspect Part 97, we find that part of the reason for the Amateur Radio Service is for "advancing skills in both the communication and technical phases of the art." This makes **Answer C** the correct choice. The Amateur Radio Service requires licensing and some skills. Answer A is not a good choice because it would let anyone have access to the Amateur Radio Service. Be careful with Answer B. While the Amateur Radio Service is non-profit and helps promote international good will, Part 97 does not contain this phrasing. Because Answers A and B are incorrect, Answer D must also be incorrect.

T1A02 [97.1] Which agency regulates and enforces the rules for the Amateur Radio Service in the United States?

A. FEMA
B. Homeland Security
C. The FCC
D. All of these choices are correct

Going back to Part 97, we find that

> In ITU Region 2, amateur radio is regulated by the FCC within the territorial limits of the 50 United States, District of Columbia, Caribbean Insular areas [Commonwealth of Puerto Rico, United States Virgin Islands (50 islets and cays) and Navassa Island], and Johnston Island (Islets East, Johnston, North and Sand) and Midway Island (Islets Eastern and Sand) in the Pacific Insular areas.

Answer C the correct choice. The other government agencies are not involved with enforcing radio regulations, so they are incorrect choices.

T1A03 [97.119(b)(2)] What are the FCC rules regarding the use of a phonetic alphabet for station identification in the Amateur Radio Service?

A. It is required when transmitting emergency messages
B. It is prohibited
C. It is required when in contact with foreign stations
D. It is encouraged

The FCC Part 97 rules state that the use "of a phonetic alphabet as an aid for correct station identification is encouraged." **Answer D** is the right choice. The other choices do not match Part 97, so they are incorrect.

T1A04 [97.5(b)(1)] How many operator/primary station license grants may be held by any one person?

A. One
B. No more than two
C. One for each band on which the person plans to operate
D. One for each permanent station location from which the person plans to operate

Part 97 states that an operator may hold "One, but only one, operator/primary station license grant may be held by any one person." This makes **Answer A** the right choice. The other options violate Part 97, so they are incorrect.

T1A05 [97.7] What is proof of possession of an FCC-issued operator/primary license grant?

A. A printed operator/primary station license issued by the FCC must be displayed at the transmitter site
B. The control operator must have an operator/primary station license in his or her possession when in control of a transmitter
C. The control operator's operator/primary station license must appear in the FCC ULS consolidated licensee database
D. All of these choices are correct

The FCC recognizes an amateur whose "operator/primary station license grant appears on the ULS consolidated licensee database." Only **Answer C** matches this Part 97 rule. The other choices are to distract you.

T1A06 [97.3(a)(9)] What is the FCC Part 97 definition of a beacon?

A. A government transmitter marking the amateur radio band edges
B. A bulletin sent by the FCC to announce a national emergency
C. An amateur station transmitting communications for the purposes of observing propagation or related experimental activities
D. A continuous transmission of weather information authorized in the amateur bands by the National Weather Service

Part 97 defines a beacon as "An amateur station transmitting communications for the purposes of observation of propagation and reception or other related experimental activities." **Answer C** matches this definition. The other choice might be nice to have, if they existed, but they are not beacon stations.

T1A07 [97.3(a)(41)] What is the FCC Part 97 definition of a space station?

A. Any satellite orbiting the earth
B. A manned satellite orbiting the earth
C. An amateur station located more than 50 km above the Earth's surface
D. An amateur station using amateur radio satellites for relay of signals

As in **Answer C**, Part 97 defines a space station as "An amateur station located more than 50 km above the Earth's surface." The other choices are to distract you.

T1A08 [97.3(a)(22)] Which of the following entities recommends transmit/receive channels and other parameters for auxiliary and repeater stations?

A. Frequency Spectrum Manager appointed by the FCC
B. Volunteer Frequency Coordinator recognized by local amateurs
C. FCC Regional Field Office
D. International Telecommunications Union

The FCC defines a Frequency Coordinator as

> An entity, recognized in a local or regional area by amateur operators whose stations are eligible to be auxiliary or repeater stations, that recommends transmit/receive channels and associated operating and technical parameters for such stations in order to avoid or minimize potential interference.

This makes **Answer B** the right choice. Be careful here because many facilities have a Frequency Spectrum Manager who does a similar function, but not for auxiliary and repeater stations, so Answer A is incorrect. It may surprise you that a FCC regional office does not perform this function, but the FCC lets the Amateur Radio Service community perform many routine functions for self-regulation making Answer C incorrect. The ITU does not operate in the US, so Answer D is also incorrect.

T1A09 [97.3(a)(22)] Who selects a Frequency Coordinator?

A. The FCC Office of Spectrum Management and Coordination Policy
B. The local chapter of the Office of National Council of Independent Frequency Coordinators
C. Amateur operators in a local or regional area whose stations are eligible to be auxiliary or repeater stations
D. FCC Regional Field Office

The discussion in the previous question will help you here. Based on the Frequency Coordinator definition, we can see that **Answer C** is the right choice. Because the

FCC has passed this function to the Amateur Radio Service community, Answers A and D cannot be correct. While the National Frequency Coordinator's Council exists, the organization given in Answer B does not exist, so this is incorrect.

T1A10 [97.3(a)(38), 97.407] Which of the following describes the Radio Amateur Civil Emergency Service (RACES)?

A. A radio service using amateur frequencies for emergency management or civil defense communications
B. A radio service using amateur stations for emergency management or civil defense communications
C. An emergency service using amateur operators certified by a civil defense organization as being enrolled in that organization
D. All of these choices are correct

The FCC defines RACES as "A radio service using amateur stations for civil defense communications during periods of local, regional or national civil emergencies." Additionally, Part 97 requires that

> No station may transmit in RACES unless it is an FCC-licensed primary, club, or military recreation station and it is certified by a civil defense organization as registered with that organization. No person may be the control operator of an amateur station transmitting in RACES unless that person holds a FCC-issued amateur operator license and is certified by a civil defense organization as enrolled in that organization.

Answers A, B, and C all contain elements of RACES, so the best choice is **Answer D.**

T1A11 [97.101 (d)] When is willful interference to other amateur radio stations permitted?

A. To stop another amateur station which is breaking the FCC rules
B. At no time
C. When making short test transmissions
D. At any time, stations in the Amateur Radio Service are not protected from willful interference

The FCC is specific about not causing harmful interference. Part 97 states that no "amateur operator shall willfully or maliciously interfere with or cause interference to any radio communication or signal." Only **Answer B** conforms with Part 97. The other choices are to distract you.

1.3 T1B – Authorized Frequencies

1.3.1 Overview

The *Authorized Frequencies* question group in Subelement T1 covers radio frequencies and emissions used the Amateur Radio Service as specified in the FCC rules. The

Table 1.1: Part 97 Common Bands and Frequencies for Technician Use in the Amateur Radio Service in Region 2 (CW allowed if not specifically listed)

Band	Frequency	Wavelength	Modes
HF	3.525–3.600 MHz	80 m	CW
HF	7.025–7.125 MHz	40 m	CW
HF	21.02–21.200 MHz	15 m	CW
HF	28.0–28.5 MHz	10 m	RTTY, data, phone
VHF	50–54 MHz	6 m	RTTY, data, phone, image
VHF	144–148 MHz	2 m	RTTY, data, phone, image
VHF	219–220 MHz	1.25 m	Fixed digital messaging
	222–225 MHz		RTTY, data, phone, image
UHF	420–450 MHz	70 cm	RTTY, data, phone, image
UHF	902–928 MHz	33 cm	RTTY, data, phone, image
UHF	1240–1300 MHz	23 cm	RTTY, data, phone, image
UHF	2300–2310 MHz	13 cm	All modes
	2390–2450 MHz		
SHF	3.3–3.5 GHz	9 cm	All modes

Authorized Frequencies questions go into

- Frequency allocations
- ITU
- Emission modes
- Restricted sub-bands
- Spectrum sharing
- Transmissions near band edges
- Contacting the International Space Station
- Power output

There is a total of 12 questions in this group of which one will be selected for the exam.

Frequency and Wavelength To answer many of the questions in this group, you will need to be able to go back and forth quickly in thinking about the Amateur Radio Service bands. One way to represent them is by their frequency designation and the other is by their wavelength designation. Table 1.1 can help with this conversion. This conversion is based on the physics relationship between the operating frequency, f, in Hz, the wavelength, λ, in meters, and the speed of light, c. The relationship is $c = \lambda f$. When we consider the operating frequency in MHz (1 000 000 Hz) and approximate the speed of light by 300 000 000 m/s, then the relationship between operating frequency and wavelength can be written as $\lambda \approx 300/f(in\ \text{MHz})$ to give the operating wavelength in meters.

Table 1.2: Transmission Modes for Operating in the Amateur Radio Service

Mode	Meaning
CW	International Morse code telegraphy emissions
Data	Telemetry, telecommand and computer communications emissions
Image	Facsimile and television emissions
MCW	Tone-modulated international Morse code telegraphy emissions
Phone	Speech and other sound emissions
RTTY	Narrow-band direct-printing telegraphy emissions
SS	Spread spectrum emissions using bandwidth-expansion modulation emissions
Test	Emissions containing no information

Transmission Modes Another type of telecommunications knowledge you will need in this section are the transmission modes. Table 1.2 lists the common modes with their names from Part 97.

1.3.2 Questions

T1B01 What is the International Telecommunications Union (ITU)?

A. An agency of the United States Department of Telecommunications Management
B. A United Nations agency for information and communication technology issues
C. An independent frequency coordination agency
D. A department of the FCC

As we saw earlier, the ITU is an international organization, so this immediately makes Answers A and D incorrect. You should be able to spot **Answer B** as correct because the ITU is a United Nations agency. The ITU is involved with frequency allocation, but does not directly coordinate, so Answer C is also incorrect.

T1B02 [97.301, 97.207(c)] Which amateur radio stations may make contact with an amateur radio station on the International Space Station (ISS) using 2 meter and 70 cm band frequencies?

A. Only members of amateur radio clubs at NASA facilities
B. Any amateur holding a Technician or higher-class license
C. Only the astronaut's family members who are hams
D. Contacts with the ISS are not permitted on amateur radio frequencies regulations

Once you have your Technician Class license, the Part 97 rules permit owning and operating an amateur space station. This is the only operational restriction, so you are eligible to make contacts with the International Space Station (ISS), and Answer B is the correct choice. Answers A and C are incorrect – NASA encourages the astronauts to interact with the general public. Answer D is technically incorrect.

T1B03 [97.301(a)] Which frequency is within the 6 meter band?

A. 49.00 MHz
B. 52.525 MHz
C. 28.50 MHz
D. 222.15 MHz

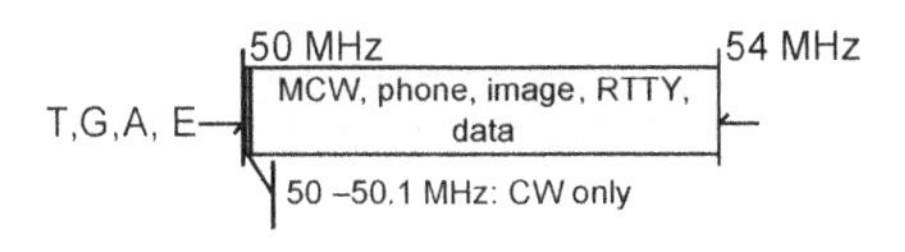

Figure 1.2: The 6-m/50-MHz band.

If we use Table 1.1 or Figure 1.2, we can see that **Answer B** is correct. As a check, the formula, 300/52.525 gives 5.7 m, which we round up to 6 m. Answer A is not a permitted Amateur Radio Service band, so it is incorrect. Answer C is in the 10-m band, while Answer D is in the 1.25-m band.

T1B04 [97.301(a)] Which amateur band are you using when your station is transmitting on 146.52 MHz?

A. 2 meter band
B. 20 meter band
C. 14 meter band
D. 6 meter band

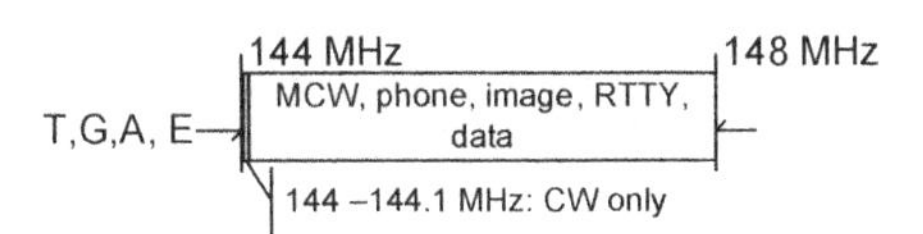

Figure 1.3: The 2-m/144-MHz band.

Again, using Table 1.1 or Figure 1.3, we can see that **Answer A** is correct. The formula, 300/146.52 gives 2 m. From the previous question, we know that Answer D is incorrect. Answer C is also incorrect because the 14-m band is not an Amateur Radio Service band. Be careful with Answer B because the 20-m band is a popular Amateur Radio Service band, but the wavelength is off by a factor of 10.

T1B05 [97.305(c)] What is the limitation for emissions on the frequencies between 219 and 220 MHz?

A. Spread spectrum only
B. Fixed digital message forwarding systems only
C. Emergency traffic only
D. Fast-scan television only

Using Table 1.1 as a guide, we can see that this segment of the 70-cm band only allows fixed digital message forwarding systems, so **Answer B** is the correct choice. The other choices are not allowed by Part 97.

T1B06 [97.301(e), 97.305] On which HF bands does a Technician class operator have phone privileges?

A. None
B. 10 meter band only
C. 80 meter, 40 meter, 15 meter and 10 meter bands
D. 30 meter band only

Going back to Table 1.1, we can see that the 10-m band is the only High Frequency (HF) band with Technician phone privileges making **Answer B** correct. Be careful with Answer C because it includes 10 m, but it also has other bands where Part 97 does not permit phone for Technicians. The other choices violate Part 97.

T1B07 [97.305(a), (c)] Which of the following VHF/UHF frequency ranges are limited to CW only?

A. 50.0 MHz to 50.1 MHz and 144.0 MHz to 144.1 MHz
B. 219 MHz to 220 MHz and 420.0 MHz to 420.1 MHz
C. 902.0 MHz to 902.1 MHZ
D. All of these choices are correct

If you refer to Figures 1.2 and 1.3, we can see that Part 97 reserves the lower band for Continuous Wave (CW) only transmissions, as in **Answer A**. Part 97 does not show that restriction for the 1.25-m, 70-cm, and 33-cm bands.

T1B08 [97.303] Which of the following is a result of the fact that the Amateur Radio Service is secondary in all or portions of some amateur bands (such as portions of the 70 cm band)?

A. U.S. amateurs may find non-amateur stations in those portions, and must avoid interfering with them
B. U.S. amateurs must give foreign amateur stations priority in those portions
C. International communications are not permitted in those portions
D. Digital transmissions are not permitted in those portions

As noted in Part 97, "A station in a secondary service must not cause harmful interference to, and must accept interference from, stations in a primary service." Therefore, **Answer A** is the correct choice. Each of the reasons given in the other answers is to see if you have read Part 97 because they make totally incorrect statements.

T1B09 [97.101(a), 97.301(a-e)] Why should you not set your transmit frequency to be exactly at the edge of an amateur band or sub-band?

A. To allow for calibration error in the transmitter frequency display
B. So that modulation sidebands do not extend beyond the band edge
C. To allow for transmitter frequency drift
D. All of these choices are correct

This will take a bit of reasoning to figure out. First, we start with the engineering principle found in Part 97: "In all respects not specifically covered by FCC Rules each amateur station must be operated in accordance with good engineering and good amateur practice." Then we look at the band allocations and notice that they have fixed edges where the amateur can operate. You need to know the amount of frequency spectrum your transmission occupies, which is always some finite number. So, to keep your signal from spilling outside the band, you cannot operate exactly on the band edge. You must also leave some allowance for your equipment not operating in perfect order, and the frequency you think you are using may be slightly different due to the equipment not being perfect. Therefore, each of the reasons given in Answers A, B, and C must be taken into account individually. That makes **Answer D** the best choice.

T1B10 [97.301(e), 97.305(c)] Which of the following HF bands have frequencies available to the Technician class operator for RTTY and data transmissions?

A. 10 meter, 12 meter, 17 meter, and 40 meter bands
B. 10 meter, 15 meter, 40 meter, and 80 meter bands
C. 30 meter band only
D. 10 meter band only

Here we need to refer to Table 1.1 again where we see that only the 10-m band permits these modes, as in **Answer D**. The other choices violate Part 97.

T1B11 [97.313] What is the maximum peak envelope power output for Technician class operators using their assigned portions of the HF bands?

A. 200 watts
B. 100 watts
C. 50 watts
D. 10 watts

In general, Part 97 requires that "An amateur station must use the minimum transmitter power necessary to carry out the desired communications." In this case, Part 97 restricts operations so that "No station may transmit with a transmitter power output exceeding 200 W PEP." Therefore, **Answer A** is the right choice. The others do not match Part 97.

T1B12 [97.313(b)] Except for some specific restrictions, what is the maximum peak envelope power output for Technician class operators using frequencies above 30 MHz?

A. 50 watts
B. 100 watts
C. 500 watts
D. 1500 watts

In this case, Part 97 states that "No station may transmit with a transmitter power exceeding 1.5 kW PEP." **Answer D** is the right choice. The other choices are to see if you have looked at the Part 97 regulations.

1.4 T1C – Operator Licensing

1.4.1 Overview

The *Operator Licensing* question group in Subelement T1 covers operator licensing in the Amateur Radio Service as described in the FCC rules. The *Operator Licensing* questions go into

- Operator classes
- Sequential and vanity call sign systems
- International communications
- Reciprocal operation
- Places where the Amateur Radio Service is regulated by the FCC
- Name and address on FCC license database
- License term
- Renewal
- Grace period

There is a total of 11 questions in this group of which one will be selected for the exam.

Call Sign System One major part of this group is the amateur call sign system. Each operating amateur must operate under a valid call sign as shown on their license. The FCC's call sign systems are:

Sequential call sign system The FCC selects the call sign from an alphabetized list corresponding to the geographic region of the licensee's mailing address and operator class.

Vanity call sign system The FCC selects the call sign from a list of available call signs requested by the licensee.

Special event call sign system The licensee selects the call sign from a list of call signs that the amateur station special event call sign data base coordinators coordinate, maintain, and disseminate.

The amateur call sign contains three parts: the prefix, the region, and the suffix. The prefix is either one or two letters (no numbers) long. One-letter prefixes must begin with K, N, or W, while two-letter prefixes are AA – AL, KA – KZ, NA – NZ, or WA – WZ. The region is a number from 0 through 9 corresponding to your FCC region. The suffix is one, two, or three letters long (no numbers). Table 1.3 shows the call sign formats for the various license classes. The FCC will generate your first license call sign through the sequential system. It will look like "KC5SJO." As you achieve higher license classes, you can elect to receive a "2x2" call sign that has a two-letter prefix and suffix, for example "AC5RI." Eventually, you can obtain a vanity call sign that has your initials or some other distinguishing characteristic, for example "NM4SH." That requires a specific application and a special fee.

Table 1.3: Amateur Radio Service Sequential License Formats

License Class	Prefix	Suffix
Amateur Extra	K, N, or W	2 letters
	2 letters starting with A, K, N, or W	1 letter
	2 letters starting with A	2 letters
General or Technician	2 letters starting with K, N, or W	3 letters

Special Event Call Signs There are also amateur special event stations that operate under special rules for a short, specified time. In this case, the call sign must have the single letter prefix K, N or W, followed by a single numeral 0 through 9, followed by a single letter A through W or Y or Z (for example, K4W, also known as a "1x1" format). The operator substitutes the special event call sign for the station's usual call sign when the station is transmitting.

1.4.2 Questions

T1C01 [97.9(a), 97.17(a)] For which licenses classes are new licenses currently available from the FCC?

A. Novice, Technician, General, Advanced
B. Technician, Technician Plus, General, Advanced
C. Novice, Technician Plus, General, Advanced
D. Technician, General, Amateur Extra

Be careful with this question. Part 97 says that the Amateur Radio Service license classes are Novice, Technician, General, Advanced, and Amateur Extra. However, Part 97 also states that "No new license grant will be issued for a Novice or Advanced Class operator/primary station." By combining both statements, we can see that **Answer D** is the only right choice for *new* license grants. The other answers have license classes that the FCC is not issuing for new license grants.

T1C02 [97.19] Who may select a desired call sign under the vanity call sign rules?

A. Only a licensed amateur with a General or Amateur Extra class license
B. Only a licensed amateur with an Amateur Extra class license
C. Only a licensed amateur who has been licensed continuously for more than 10 years
D. Any licensed amateur

Part 97 states that "The person named in an operator/primary station license grant or in a club station license grant is eligible to make application for modification of the license grant, or the renewal thereof, to show a call sign selected by the vanity call sign system." Therefore, any licensed amateur is eligible to have a vanity call sign, as in **Answer D**. The other choices conflict with Part 97, so they are incorrect.

T1C03 [97.117] What types of international communications is an FCC-licensed amateur radio station permitted to make?

A. Communications incidental to the purposes of the Amateur Radio Service and remarks of a personal character
B. Communications incidental to conducting business or remarks of a personal nature
C. Only communications incidental to contest exchanges, all other communications are prohibited
D. Any communications that would be permitted by an international broadcast station

In Part 97, we find that "Transmissions to a different country, where permitted, shall be limited to communications incidental to the purposes of the amateur service and to remarks of a personal character." **Answer A** is directly from Part 97, so it is the right answer. You should recognize that "business" in Answer B and "broadcast" in Answer D are flags to indicate an incorrect choice. Amateurs can do more than contest exchanges, so Answer C is incorrect.

T1C04 [97.107] When are you allowed to operate your amateur station in a foreign country?

A. When the foreign country authorizes it
B. When there is a mutual agreement allowing third party communications
C. When authorization permits amateur communications in a foreign language
D. When you are communicating with non-licensed individuals in another country

The FCC and the foreign government must have a reciprocal operating agreement for licensed amateurs to operate in the other countries, so **Answer A** is the correct choice. This is actual operation, and not third-party communications, so Answer B is incorrect. Part 97 permits foreign language operation in the US, so Answer C is incorrect. A licensed amateur cannot communicate with non-licensed individual, so Answer D is also incorrect.

T1C05 Which of the following is a valid call sign for a Technician class amateur radio station?

A. K1XXX
B. KA1X
C. W1XX
D. All of these choices are correct

As we saw in Table 1.3, a Technician class license cannot have a suffix of one or two letters, so Answers B and C are incorrect. This makes Answer D incorrect too. **Answer A** is the right choice because the suffix is three letters long and the call sign begins with a K.

T1C06 [97.5(a)(2)] From which of the following locations may an FCC-licensed amateur station transmit?

A. From within any country that belongs to the International Telecommunications Union
B. From within any country that is a member of the United Nations
C. From anywhere within International Telecommunications Union (ITU) Regions 2 and 3
D. From any vessel or craft located in international waters and documented or registered in the United States

The FCC only has regulatory authority in three locations:

1 Within 50 km of the Earth's surface and at a place where the amateur service is regulated by the FCC
2 Within 50 km of the Earth's surface and aboard any vessel or craft that is documented or registered in the United States
3 More than 50 km above the Earth's surface aboard any craft that is documented or registered in the United States

Answer D is the only location listed that totally lies within the FCC's regulatory domain. While the FCC is responsible for parts of ITU Regions 2 and 3, it is not responsible for all of it, so Answer C is incorrect. The others are to distract you.

T1C07 [97.23] What may result when correspondence from the FCC is returned as undeliverable because the grantee failed to provide and maintain a correct mailing address with the FCC?

A. Fine or imprisonment
B. Revocation of the station license or suspension of the operator license
C. Require the licensee to be re-examined
D. A reduction of one rank in operator class

As in **Answer B**, Part 97 states that "Revocation of the station license or suspension of the operator license may result when correspondence from the FCC is returned as undeliverable because the grantee failed to provide the correct mailing address." The other choices do not agree with Part 97, so they are incorrect.

T1C08 [97.25] What is the normal term for an FCC-issued primary station/operator amateur radio license grant?

A. Five years
B. Life
C. Ten years
D. Twenty years

Part 97 states that "An amateur service license is normally granted for a 10-year term." This makes **Answer C** correct, and the other choices are not consistent with

Part 97.

T1C09 [97.21(a)(b)] What is the grace period following the expiration of an amateur license within which the license may be renewed?

A. Two years
B. Three years
C. Five years
D. Ten years

Part 97 states that "A person whose amateur station license grant has expired may apply to the FCC for renewal of the license grant for another term during a 2 year filing grace period." This makes **Answer A** the right choice. Be careful with Answer D because that is the term of a license grant, but it is incorrect for this question. The other choices do not match Part 97, so they are wrong.

T1C10 [97.5a] How soon after passing the examination for your first amateur radio license may you operate a transmitter on an Amateur Radio Service frequency?

A. Immediately
B. 30 days after the test date
C. As soon as your operator/station license grant appears in the FCC's license database
D. You must wait until you receive your license in the mail from the FCC

Part 97 says that the station equipment "must be under the physical control of a person named in an amateur station license grant on the ULS consolidated license database." This means that as soon as the license appears in the Universal Licensing System (ULS), the licensee is authorized to operate, as indicated in **Answer C**. Answer D was true many years ago, but not now with the Web. Answers A and B violate Part 97, so they are incorrect.

T1C11 [97.21(b)] If your license has expired and is still within the allowable grace period, may you continue to operate a transmitter on Amateur Radio Service frequencies?

A. No, transmitting is not allowed until the FCC license database shows that the license has been renewed
B. Yes, but only if you identify using the suffix GP
C. Yes, but only during authorized nets
D. Yes, for up to two years

Part 97 states that during the grace period, "Unless and until the license grant is renewed, no privileges in this part are conferred." That means no operating privileges during the grace period, and **Answer A** is the right choice. The other choices violate Part 97 in allowing operating privileges, so they are incorrect.

1.5 T1D – Authorized and Prohibited Transmission

Table 1.4: Station Types

Type	Definition
Amateur station	A station in an amateur radio service consisting of the apparatus necessary for carrying on radio communications
Auxiliary station	An amateur station, other than in a message forwarding system, that is transmitting communications point-to-point within a system of cooperating amateur stations
Beacon	An amateur station transmitting communications for the purposes of observation of propagation and reception or other related experimental activities
Earth station	An amateur station located on, or within 50 km of, the Earth's surface intended for communications with space stations or with other Earth stations by means of one or more other objects in space
Message forwarding system	A group of amateur stations participating in a voluntary, cooperative, interactive arrangement where communications are sent from the control operator of an originating station to the control operator of one or more destination stations by one or more forwarding stations
Repeater	An amateur station that simultaneously retransmits the transmission of another amateur station on a different channel or channels
Space station	An amateur station located more than 50 km above the Earth's surface
Telecommand station	An amateur station that transmits communications to initiate, modify or terminate functions of a space station

1.5.1 Overview

The *Authorized and Prohibited Transmission* question group in Subelement T1 covers aspects of permitted transmissions in the Amateur Radio Service described in the FCC rules. The *Authorized and Prohibited Transmission* questions go into

- Communications with other countries
- Music
- Exchange of information with other services
- Indecent language
- Compensation for use of station
- Retransmission of other amateur signals
- Codes and ciphers

- Sale of equipment
- Unidentified transmissions
- Oner-way transmission

There is a total of 11 questions in this group of which one will be selected for the exam.

Station Types This section covers some different station types found in amateur radio. Besides your own station, there are repeaters, beacons, and other ways of exchanging information. Table 1.4 lists the formal definitions for various types of stations authorized with amateur radio.

1.5.2 Questions

T1D01 [97.111(a)(1)] With which countries are FCC-licensed amateur stations prohibited from exchanging communications?

A. Any country whose administration has notified the International Telecommunications Union (ITU) that it objects to such communications
B. Any country whose administration has notified the American Radio Relay League (ARRL) that it objects to such communications
C. Any country engaged in hostilities with another country
D. Any country in violation of the War Powers Act of 1934

Under Part 97, amateurs may exchange communications with foreign stations "in the amateur service, except those in any country whose administration has notified the ITU that it objects to such communications." This makes **Answer A** the right choice. While the ARRL of Answer B may publish a list of such countries for amateur convenience, it is not official, so this is incorrect. Answers C and D may sound official, but they are to distract you.

T1D02 [97.113(b),97.111(b)] Under which of the following circumstances may an amateur radio station make one-way transmissions?

A. Under no circumstances
B. When transmitting code practice, information bulletins, or transmissions necessary to provide emergency communications
C. At any time, as long as no music is transmitted
D. At any time, as long as the material being transmitted did not originate from a commercial broadcast station

Part 97 permits one-way transmissions for the purposes of

(1) Brief transmissions necessary to make adjustments to the station;
(2) Brief transmissions necessary to establishing two-way communications with other stations;
(3) Telecommand;

(4) Transmissions necessary to providing emergency communications;
(5) Transmissions necessary to assisting persons learning, or improving proficiency in, the international Morse code; and
(6) Transmissions necessary to disseminate information bulletins.
(7) Transmissions of telemetry

Of the choices given, only **Answer B** matches part of this list, so it is the right choice. The other choices are not in Part 97, so they are incorrect.

T1D03 [97.211(b), 97.215(b), 97.114(a)(4)] When is it permissible to transmit messages encoded to hide their meaning?

A. Only during contests
B. Only when operating mobile
C. Only when transmitting control commands to space stations or radio control craft
D. Only when frequencies above 1280 MHz are used

Generally, amateur radio transmissions are to be open and available to all listeners. However, there are important exceptions to protect certain station classes. The first is for a telecommand station that "may transmit special codes intended to obscure the meaning of telecommand messages to the station in space operation." The second is for controlling model aircraft where the "control signals are not considered codes or ciphers intended to obscure the meaning of the communication." From this, we see that **Answer C** is the correct choice. The other answers do not match Part 97.

T1D04 [97.113(a)(4), 97.113(c)] Under what conditions is an amateur station authorized to transmit music using a phone emission?

A. When incidental to an authorized retransmission of manned spacecraft communications
B. When the music produces no spurious emissions
C. When the purpose is to interfere with an illegal transmission
D. When the music is transmitted above 1280 MHz

Generally, an amateur station may not transmit music, including musical tones from a computer, except "music, originating on United States Government frequencies between a manned spacecraft and its associated Earth stations." This makes **Answer A** the right choice. The other choices are not in Part 97, so they are incorrect.

T1D05 [97.113(a)(3)(ii)] When may amateur radio operators use their stations to notify other amateurs of the availability of equipment for sale or trade?

A. When the equipment is normally used in an amateur station and such activity is not conducted on a regular basis
B. When the asking price is $100.00 or less
C. When the asking price is less than its appraised value
D. When the equipment is not the personal property of either the station licensee or the control operator or their close relatives

While the FCC does not permit amateurs to use the service as a business, the FCC does make a small exception for selling amateur equipment on a limited basis. Part 97 says that an amateur operator "may notify other amateur operators of the availability for sale or trade of apparatus normally used in an amateur station, provided that such activity is not conducted on a regular basis." This makes **Answer A** the right choice. The others do not match Part 97, so they are incorrect.

T1D06 [97.113(a)(4)] What, if any, are the restrictions concerning transmission of language that may be considered indecent or obscene?

A. The FCC maintains a list of words that are not permitted to be used on amateur frequencies
B. Any such language is prohibited
C. The ITU maintains a list of words that are not permitted to be used on amateur frequencies
D. There is no such prohibition

Neither the FCC nor the ITU has a banned-words list, so Answers A and C are incorrect. Part 97 says that amateurs may not transmit "obscene or indecent words or language" making **Answer B** the right choice. Answer D is a violation of Part 97.

T1D07 [97.113(d)] What types of amateur stations can automatically retransmit the signals of other amateur stations?

A. Auxiliary, beacon, or Earth stations
B. Repeater, auxiliary, or space stations
C. Beacon, repeater, or space stations
D. Earth, repeater, or space stations

In Part 97, we see that "No amateur station, except an auxiliary, repeater, or space station, may automatically retransmit the radio signals of other amateur station." This makes **Answer B** the right choice. Be careful with the other choices because they contain a mix of permitted and non-permitted station types, so you must read the choices carefully during the examination.

T1D08 [97.113(a)(3)(iii)] In which of the following circumstances may the control operator of an amateur station receive compensation for operating the station?

A. When the communication is related to the sale of amateur equipment by the control operator's employer
B. When the communication is incidental to classroom instruction at an educational institution
C. When the communication is made to obtain emergency information for a local broadcast station
D. All of these choices are correct

Part 97 permits a control operator to "accept compensation as an incident of a teaching position during periods of time when an amateur station is used by that teacher as a part of classroom instruction at an educational institution." This makes **Answer B** the correct choice. Answers A and C are inconsistent with Part 97, so they are incorrect. This also makes Answer D incorrect.

T1D09 [97.113(5)(b)] Under which of the following circumstances are amateur stations authorized to transmit signals related to broadcasting, program production, or news gathering, assuming no other means is available?

A. Only where such communications directly relate to the immediate safety of human life or protection of property
B. Only when broadcasting communications to or from the space shuttle
C. Only where noncommercial programming is gathered and supplied exclusively to the National Public Radio network
D. Only when using amateur repeaters linked to the internet

Generally, amateur stations may not engage in broadcasting (transmitting to the general public) except

> that communications directly related to the immediate safety of human life or the protection of property may be provided by amateur stations to broadcasters for dissemination to the public where no other means of communication is reasonably available before or at the time of the event.

This makes **Answer A** the correct choice. Be careful with Answer B because amateurs can retransmit space communications, but not broadcast them, so this is an incorrect choice. Answers C and D are to distract you.

T1D10 [97.3(a)(10)] What is the meaning of the term broadcasting in the FCC rules for the amateur services?

A. Two-way transmissions by amateur stations
B. Transmission of music
C. Transmission of messages directed only to amateur operators
D. Transmissions intended for reception by the general public

Part 97 defines broadcasting as "Transmissions intended for reception by the general public, either direct or relayed." This makes **Answer D** the right choice.

T1D11 [97.119(a)] When may an amateur station transmit without on-the-air identification?

A. When the transmissions are of a brief nature to make station adjustments
B. When the transmissions are unmodulated
C. When the transmitted power level is below 1 watt
D. When transmitting signals to control a model craft

Part 97 states that each "amateur station, except a space station or telecommand station, must transmit its assigned call sign." This makes **Answer D** the correct choice. The other answers do not conform with this requirement of Part 97.

1.6 T1E – Control Operator and Control Types

1.6.1 Overview

The *Control Operator and Control Types* question group in Subelement T1 covers station control procedures the Amateur Radio Service defined in the FCC rules. The *Control Operator and Control Types* questions go into

- Control operator required
- Eligibility
- Designation of control operator
- Privileges and duties
- Control point
- Local, automatic and remote control
- Location of control operator

There is a total of 11 questions in this group of which one will be selected for the exam.

1.6.2 Questions

T1E01 [97.7(a)] When is an amateur station permitted to transmit without a control operator?

A. When using automatic control, such as in the case of a repeater
B. When the station licensee is away and another licensed amateur is using the station
C. When the transmitting station is an auxiliary station
D. Never

Part 97 states that when "transmitting, each amateur station must have a control operator." This makes **Answer D** the correct choice. The other answers represent cases where Part 97 requires a control operator, so they are incorrect choices.

T1E02 [97.301, 97.207(c)] Who may be the control operator of a station communicating through an amateur satellite or space station?

A. Only an Amateur Extra Class operator
B. A General class or higher licensee who has a satellite operator certification
C. Only an Amateur Extra Class operator who is also an AMSAT member
D. Any amateur whose license privileges allow them to transmit on the satellite uplink frequency

Figure 1.4: A university-built small satellite.

Part 97 states that "Any amateur station may be a space station. A holder of any class operator license may be the control operator of a space station, subject to the privileges of the class of operator license held by the control operator," which makes **Answer D** the right choice for the operator's minimal operating privileges. Answers A and C are restrictions that do not exist. Answer B is wrong because, in part, there is not such endorsement in the Amateur Radio Service. Figure 1.4 shows example of an amateur satellite. If you are interested in satellites, you may wish to investigate the Radio Amateur Satellite Corporation (AMSAT) at `http://www.amsat.org`.

T1E03 [97.103(b)] Who must designate the station control operator?

A. The station licensee
B. The FCC
C. The frequency coordinator
D. The ITU

The FCC requires that the "station licensee must designate the station control operator." This makes **Answer A** the right choice.

T1E04 [97.103(b)] What determines the transmitting privileges of an amateur station?

A. The frequency authorized by the frequency coordinator
B. The frequencies printed on the license grant
C. The highest class of operator license held by anyone on the premises
D. The class of operator license held by the control operator

Part 97 indicates that the control operator's license class determines the transmitting privileges, so **Answer D** is the correct choice. The other choices are to distract you.

T1E05 [97.3(a)(14)] What is an amateur station control point?

A. The location of the station's transmitting antenna
B. The location of the station transmitting apparatus
C. The location at which the control operator function is performed
D. The mailing address of the station licensee

Part 97 defines control point as "The location at which the control operator function is performed." This makes **Answer C** the correct choice. The other locations are all important in the process, but they do not match the definition in Part 97, so they are incorrect.

T1E06 [97.119(e)] When, under normal circumstances, may a Technician class licensee be the control operator of a station operating in an exclusive Amateur Extra class operator segment of the amateur bands?

A. At no time
B. When operating a special event station
C. As part of a multi-operator contest team
D. When using a club station whose trustee is an Amateur Extra class operator licensee

Part 97 permits operation when "the operator license class held by the control operator exceeds that of the station licensee, an indicator consisting of the call sign assigned to the control operator's station must be included after the call sign." This case is the inverse of that, so the right choice is **Answer A**. In the other choices, the Technician operator would not be operating under their own license grant, but that of the mentioned station, so they are not correct choices here.

T1E07 [97.103(a)] When the control operator is not the station licensee, who is responsible for the proper operation of the station?

A. All licensed amateurs who are present at the operation
B. Only the station licensee
C. Only the control operator
D. The control operator and the station licensee are equally responsible

Part 97 says that "When the control operator is a different amateur operator than the station licensee, both persons are equally responsible for proper operation of the station." This makes **Answer D** the right choice for this question. Answers B and C are incomplete, so they are incorrect. Answer A may not include the relevant persons, so it is also incorrect.

T1E08 [97.3(a)(6), 97.205(d)] Which of the following is an example of automatic control?

A. Repeater operation
B. Controlling the station over the internet
C. Using a computer or other device to automatically send CW
D. Using a computer or other device to automatically identify

We start with the definition in Part 97, which says automatic control is "The use of devices and procedures for control of a station when it is transmitting so that compliance with the FCC Rules is achieved without the control operator being present at a control point." Under Part 97, we learn that a "repeater may be automatically controlled." This makes **Answer A** the best choice. The other choices are not specific enough nor are they specifically called out in Part 97, so they are not good options for this question.

T1E09 [97.109(c)] Which of the following is true of remote control operation?
A. The control operator must be at the control point
B. A control operator is required at all times
C. The control operator indirectly manipulates the controls
D. All of these choices are correct

Each of the elements of Answers A, B, and C are involved with remote control operations. Therefore, **Answer D** is the best choice.

T1E10 [97.3(a)(39)] Which of the following is an example of remote control as defined in Part 97?
A. Repeater operation
B. Operating the station over the internet
C. Controlling a model aircraft, boat or car by amateur radio
D. All of these choices are correct

Part 97 defines remote control as "The use of a control operator who indirectly manipulates the operating adjustments in the station through a control link to achieve compliance with the FCC Rules." **Answer B** is an example of remote control, and is the best choice to answer this question. Be careful with your reading of this question because remote control implies using something other than direct control via an amateur radio link, so Answer C is incorrect. Because of this, Answer D is also incorrect. Answer A is to distract you.

T1E11 [97.103(a)] Who does the FCC presume to be the control operator of an amateur station, unless documentation to the contrary is in the station records?
A. The station custodian
B. The third party participant
C. The person operating the station equipment
D. The station licensee

Part 97 states that "The FCC will presume that the station licensee is also the control operator, unless documentation to the contrary is in the station records." Therefore, the station licensee, as in **Answer D**, is the right choice. You should be able to spot the other choices as distraction answers.

1.7 T1F – Station Operations

1.7.1 Overview

The *Station Operations* question group in Subelement T1 covers aspects of how operators identify stations according to the FCC rules. The *Station Operations* questions go into

- Station identification

- Repeaters
- Third party communications
- Club stations
- FCC inspection

There is a total of 11 questions in this group of which one will be selected for the exam. This section revisits some different station types, so you may wish to refresh your look at Table 1.4.

1.7.2 Questions

T1F1 [97.103(c)] When must the station licensee make the station and its records available for FCC inspection?

A. At any time ten days after notification by the FCC of such an inspection
B. At any time upon request by an FCC representative
C. Only after failing to comply with an FCC notice of violation
D. Only when presented with a valid warrant by an FCC official or government agent

Part 97 is very clear on this point: "The station licensee must make the station and the station records available for inspection upon request by an FCC representative." This makes **Answer B** the correct choice, and the other choices are not consistent with Part 97.

T1F02 [97.119 (a)] When using tactical identifiers such as "Race Headquarters" during a community service net operation, how often must your station transmit the station's FCC-assigned call sign?

A. Never, the tactical call is sufficient
B. Once during every hour
C. At the end of each communication and every ten minutes during a communication
D. At the end of every transmission

Part 97 requires that

> Each amateur station, except a space station or telecommand station, must transmit its assigned call sign on its transmitting channel at the end of each communication, and at least every 10 minutes during a communication, for the purpose of clearly making the source of the transmissions from the station known to those receiving the transmissions. No station may transmit unidentified communications or signals, or transmit as the station call sign, any call sign not authorized to the station.

This makes **Answer C** the correct choice. Be careful with Answer D because it is half right, but not complete. The other answers are to distract you.

T1F03 [97.119(a)] When is an amateur station required to transmit its assigned call sign?

A. At the beginning of each contact, and every 10 minutes thereafter
B. At least once during each transmission
C. At least every 15 minutes during and at the end of a communication
D. At least every 10 minutes during and at the end of a communication

Based on the previous question, we should be able to spot **Answer D** as the right choice. Notice that the call sign identification rule is the same for all transmission types except for space stations and telecommand stations.

T1F04 [97.119(b)(2)] Which of the following is an acceptable language to use for station identification when operating in a phone sub-band?

A. Any language recognized by the United Nations
B. Any language recognized by the ITU
C. The English language
D. English, French, or Spanish

Depending on the type of transmission, the call sign transmission rules are

(1) By a CW emission. When keyed by an automatic device used only for identification, the speed must not exceed 20 words per minute.
(2) By a phone emission in the English language. Use of a phonetic alphabet as an aid for correct station identification is encouraged.
(3) By a RTTY emission using a specified digital code when all or part of the communications are transmitted by a RTTY or data emission.
(4) By an image emission conforming to the applicable transmission standards, either color or monochrome, of §73.682(a) of the FCC Rules when all or part of the communications are transmitted in the same image emission.

For phone transmissions, Part 97 permits only English, as in **Answer C**. Be careful with Answer D because it includes English, but it also includes other languages that Part 97 does not permit.

T1F05 [97.119(b)(2)] What method of call sign identification is required for a station transmitting phone signals?

A. Send the call sign followed by the indicator RPT
B. Send the call sign using CW or phone emission
C. Send the call sign followed by the indicator R
D. Send the call sign using only phone emission

While it is natural to use a voice method for the call sign identification when transmitting on phone, Part 97 also permits operators to use CW making **Answer B** the right choice. Be careful with Answer D because it may seem to be the logical choice, but Part 97 allows CW too, so it is not the best choice here. The other choices are distractions.

T1F06 [97.119(c)] Which of the following formats of a self-assigned indicator is acceptable when identifying using a phone transmission?

A. KL7CC stroke W3
B. KL7CC slant W3
C. KL7CC slash W3
D. All of these choices are correct

Part 97 states that "One or more indicators may be included with the call sign. Each indicator must be separated from the call sign by the slant mark (/) or by any suitable word that denotes the slant mark. If an indicator is self-assigned, it must be included before, after, or both before and after, the call sign." From this wording, Part 97 permits each of the choices in Answers A, B, and C. This makes **Answer D** the best choice for this question. This example could be for an amateur temporarily operating away from their home location and is now in FCC region 3.

T1F07 [97.115(a)(2)] Which of the following restrictions apply when a non-licensed person is allowed to speak to a foreign station using a station under the control of a Technician class control operator?

A. The person must be a U.S. citizen
B. The foreign station must be one with which the U.S. has a third-party agreement
C. The licensed control operator must do the station identification
D. All of these choices are correct

The FCC rules allow third-party communications between

(1) Any station within the jurisdiction of the United States.
(2) Any station within the jurisdiction of any foreign government when transmitting emergency or disaster relief communications and any station within the jurisdiction of any foreign government whose administration has made arrangements with the United States to allow amateur stations to be used for transmitting international communications on behalf of third parties. No station shall transmit messages for a third party to any station within the jurisdiction of any foreign government whose administration has not made such an arrangement. This prohibition does not apply to a message for any third party who is eligible to be a control operator of the station.

Based on Part 97, there must be an agreement between the US and the foreign government for the foreign station, which makes **Answer B** the correct choice. Since Answers A and C are not in the Part 97 wording, they are incorrect. This also makes Answer D incorrect.

T1F08 [97.3(a)(47)] What is meant by the term Third Party Communications?

A. A message from a control operator to another amateur station control operator on behalf of another person
B. Amateur radio communications where three stations are in communications with one another
C. Operation when the transmitting equipment is licensed to a person other than the control operator
D. Temporary authorization for an unlicensed

Part 97 defines Third Party Communications as "A message from the control operator (first party) of an amateur station to another amateur station control operator (second party) on behalf of another person (third party)." **Answer A** matches this definition and the other choices are to distract you.

T1F09 [97.3(a)(40)] What type of amateur station simultaneously retransmits the signal of another amateur station on a different channel or channels?

A. Beacon station
B. Earth station
C. Repeater station
D. Message forwarding station

If we refer to Table 1.4, we can see that the repeater station of **Answer C** matches the definition in the question, so this is the right choice. If you look again at the station definitions in Table 1.4, you can see that each of the other choices is incorrect.

T1F10 [97.205(g)] Who is accountable should a repeater inadvertently retransmit communications that violate the FCC rules?

A. The control operator of the originating station
B. The control operator of the repeater
C. The owner of the repeater
D. Both the originating station and the repeater owner

One might suspect that the controller of the repeater has some level of responsibility here, but Part 97 states that "The control operator of a repeater that retransmits inadvertently communications that violate the rules in this part is not accountable for the violative communications." This makes Answer B incorrect. Notice that Part 97 does not mention the owner of the repeater, so Answers C and D are also incorrect. **Answer A** is the correct choice.

T1F11 [97.5(b)(2)] Which of the following is a requirement for the issuance of a club station license grant?

A. The trustee must have an Amateur Extra class operator license grant
B. The club must have at least four members
C. The club must be registered with the American Radio Relay League
D. All of these choices are correct

The Part 97 club requirements are:

> A club station license grant may be held only by the person who is the license trustee designated by an officer of the club. The trustee must be a person who holds an operator/primary station license grant. The club must be composed of at least four persons and must have a name, a document of organization, management, and a primary purpose devoted to amateur service activities consistent with this part.

Only **Answer B** is in these requirements. The other choices may be good ideas, but they are distractions.

Chapter 2

T2 – OPERATING PROCEDURES

2.1 Introduction

In the previous chapter, we looked at radio rules and regulations. In that chapter, most of the questions were in relation to some aspect of Part 97. In this chapter, there will be a few Part 97-related questions. However, most of the questions are based on current Amateur Radio Service practices. Since you have not operated on the air yet, some of these questions may seem to have arbitrary answers. Do not worry, these will become second nature as you gain more experience, and the normal operating procedures help the Amateur Radio Service community to keep some order on the bands and make operations fair for all users. Use the questions and the explanations to learn about accepted amateur radio practice. This will help you when you start to exercise your privileges.

This *Operating Procedures* subelement has the following question groups:

A. Station operation
B. VHF/UHF operating practices
C. Public service

This will generate three questions on the Technician examination.

2.1.1 Radio Engineering Concepts

We will need to understand radio engineering concepts to understand the intent and meaning of some of the questions in this chapter. There are some technical terms to learn, and a little mathematics to see what is going on. The diagrams should help make the math a bit easier to understand.

Repeaters Many new Technician operators use hand-held or portable rigs like those shown on the cover to make their first contacts using a local repeater. There

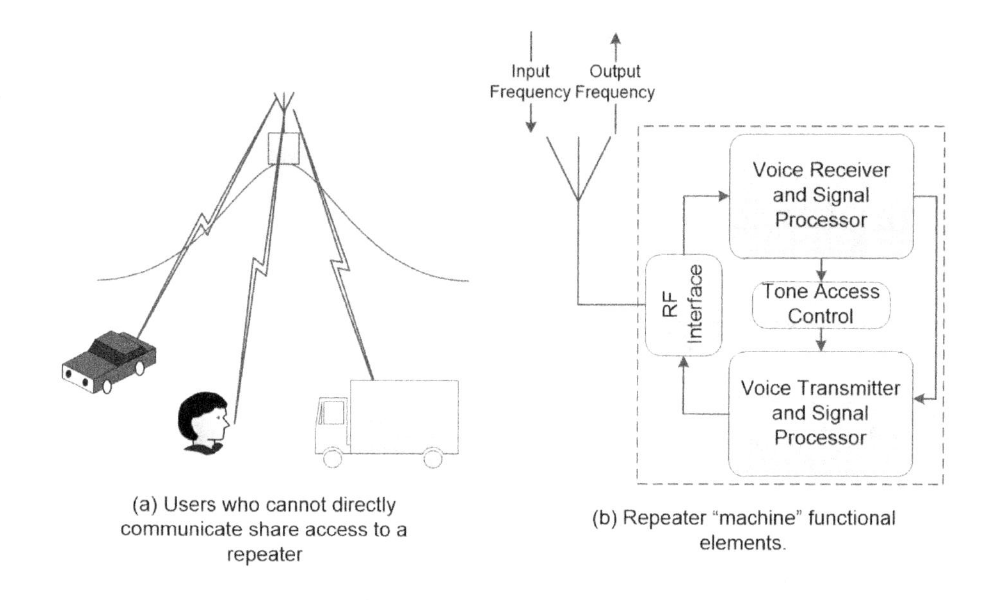

Figure 2.1: Amateur radio repeater concepts.

are several questions in this subelement to acquaint you with repeater operations. A repeater is a radio station with automatic transmission and receiving elements, and it is usually located on a hill or tall building for visibility. Figure 2.1 shows the concepts behind repeater operation. In Part (a), we see the owners placed the repeater on a hill that is visible to users in vehicles or by themselves. Part (b) shows the radio equipment configuration:

Antenna for receiving and transmitting the radio signal

Frequencies with one for incoming signals and one for outgoing signals; never the same for both

Receiver to process the incoming signals which may include an access tone

Transmitter to retransmit the incoming signal for everyone to access

The difference between the input and output frequencies is known as the *frequency offset*. The tone control in the figure keeps the repeater from responding to stray signals, and only respond to user signals preceded by the access tone.

Repeaters are generally not intended to be one-to-one communications. They are more like a gathering of friends where everyone can join in – with a little discipline to keep from stepping on toes.

Squelch Circuit A circuit that is common to both repeaters and regular radio equipment is a squelch circuit. This circuit keeps the receiver from trying to use low intensity signals that are too close to the background radio noise and too weak to understand. The operator can set the squelch threshold based on operating conditions.

Table 2.1: Selected Q Signals for the Technician Exam

Signal	Meaning
CQ	Calling any Amateur Service station
QRM	Do you have interference? [from other stations]
QRN	Are you troubled by static?
QRP	Shall I decrease power?
QRU	Have you anything for me?
QRZ	Who is calling me?
QSB	Are my signals fading?
QSL	Can you acknowledge receipt?
QSY	Shall I change to transmission on another frequency?
QTH	What is your position?

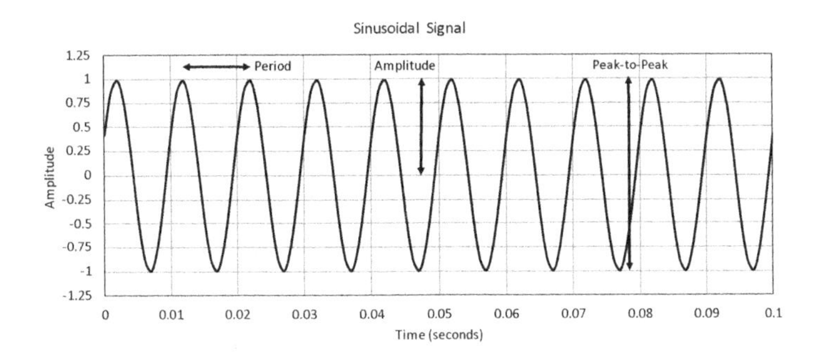

Figure 2.2: The unmodulated carrier signal used in radio systems.

Q Signals In this chapter, you will see several questions on the "Q signals" or shorthand for various standard operating conditions or modes. Table 2.1 lists some common Q signals you may wish to memorize to help with your operating techniques. You can find an extensive list at `https://en.wikipedia.org/wiki/Q_code`.

The Carrier Signal The questions in this chapter start asking about properties of radio transmissions. We will need the knowledge necessary to answer these questions again in later chapters. Here, we start with the carrier signal in describing the radio transmission. This is a sinusoidal-shaped signal that varies in time and looks like Figure 2.2. We describe this signal by its amplitude and frequency. Figure 2.2 shows two amplitude measures: the *peak amplitude* relative to the 0 level on the vertical axis, and the *peak-to-peak amplitude* that goes from the highest point in the signal to the lowest point. Note that this signal is symmetric, that is, the positive peaks are the same height as the negative peaks.

Engineers call the time between peaks the carrier's *period*. Figure 2.2 has an arrow to mark the period of the sinusoidal wave in seconds. The reciprocal of the period is the frequency in units of Hertz, or old style, in cycles per second. The speed of the carrier through the air is the speed of light or 299 792 458 m/s.

When we describe the carrier signal, $s(t)$, mathematically, we write it as either a sine or a cosine function in time, t, using

$$s(t) = A \sin 2\pi f t + \theta$$

Here, A is the carrier's amplitude, f is the carrier's frequency, and θ is the carrier's phase angle. These are important factors for understanding modulation, which comes next.

Modulation Modulation is the process where the transmitter modifies the carrier to send the message signal over the airwaves. The message signal is the item of interest to the communicators. It can be a voice transmission, text characters, numbers, the output from a temperature sensor, or whatever type of information Part 97 permits on the bands. Engineers are clever in their naming of the modulation types. They use

AM when the carrier's amplitude is proportional to the message amplitude
FM when the carrier's frequency is proportional to the message amplitude
PM when the carrier's phase angle is proportional to the message amplitude

The Amplitude Modulation (AM) signal comes in two modes: Dual Sideband (DSB) and Single Sideband (SSB). Both contain all the necessary information to convey the message. SSB uses one half of the transmission bandwidth of DSB, which makes it more popular on the bands.

The Frequency Modulation (FM) method makes the carrier change frequency moment-by-moment as the message signal changes. We write this as

$$f(t) = k_f m(t)$$

Here, $f(t)$ is the moment-by-moment carrier frequency in Hz, k_f is a scaling constant that says how strongly to react to the message signal, and $m(t)$ is the message information transmitted moment-by-moment. Figure 2.3 shows an example of a message signal, and the FM carrier frequency that results. Engineers call the frequency difference between the instantaneous carrier frequency and the unmodulated carrier frequency the *frequency deviation*.

The Phase Modulation (PM) signal makes the carrier's phase angle change moment-by-moment as the message signal changes. We write this as

$$\theta(t) = k_p m(t)$$

Here, $\theta(t)$ is the moment-by-moment carrier phase in radian units, k_p is a constant of proportionality that says how strongly to react to the message signal, and $m(t)$ is the moment-by-moment information we desire to send.

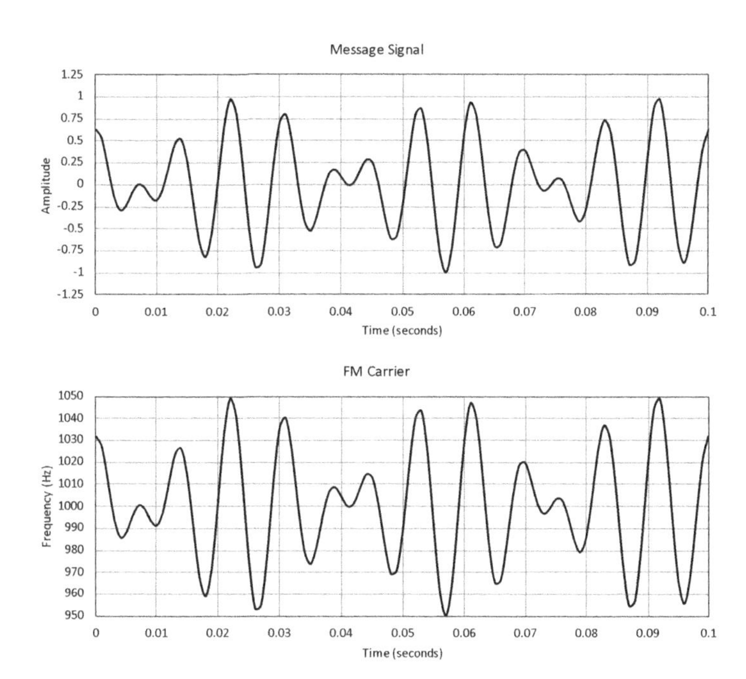

Figure 2.3: The message signal and the resulting FM carrier. The unmodulated carrier frequency is 1000 Hz.

Signal Bandwidth The radio spectrum is finite, and users must share it. The different messages and modulation formats can have different requirements for how much frequency spectrum they require for proper transmission. Understanding how these differing modulation formats work with the message signals will help in understanding how we share the bands. The most important requirement is knowing the transmitted signal's *bandwidth*. This is determined by looking at the signal with a test instrument called a *spectrum analyzer*, commonly called a *spec an*.

AM Bandwidth The AM signal has a bandwidth proportional to the bandwidth of the message signal. DSB requires twice the bandwidth of the message signal, while SSB requires the same bandwidth as the message signal. This does not change if it is a voice signal, a data signal, or a measurement signal. For example, a voice signal requires 2–3 kHz of bandwidth, while a data signal requires about 1 Hz per bit per second of data transmission (1024 bps requires about 1024 Hz of bandwidth).

FM and PM Bandwidth With FM and PM, the amount of bandwidth occupied on the radio bands depends on the amplitude of the signal, all other things being equal. For example, Figure 2.4 show the FM spectrum for the message signal from Figure 2.3. The difference in the two plots is that we have doubled the message amplitude for the second one from the first one. If you look at how wide the plots are, you can

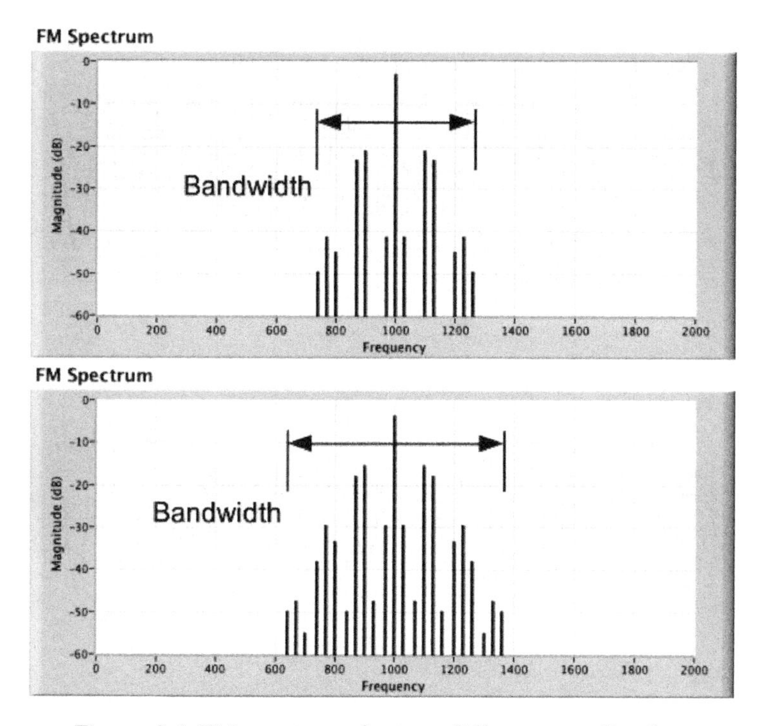

Figure 2.4: FM spectrum for two different amplitudes.

see the width of the plot along the x-axis is greater in the second. In both cases, the FM spectrum is much wider than sending the same message by DSB or SSB. For example, commercial FM broadcast uses 75 kHz to send the same information as commercial AM radio does in a few kHz.

2.2 T2A – Station Operation

2.2.1 Overview

The *Station Operation* question group in Subelement T2 introduces you to proper operating techniques in the Amateur Radio Service. The *Station Operation* group covers topics such as

- Choosing an operating frequency
- Calling another station
- Test transmissions
- Procedural signs
- Use of minimum power
- Choosing an operating frequency
- Band plans
- Calling frequencies

- Repeater offsets

There is a total of 12 questions in this group of which one will be selected for the exam.

2.2.2 Questions

T2A01 What is the most common repeater frequency offset in the 2 meter band?

A. Plus 500 kHz
B. Plus or minus 600 kHz
C. Minus 500 kHz
D. Only plus 600 kHz

As Figure 2.1 shows, repeaters need two frequencies for operation: one for the input signals and one for the output signals. Normally, the repeater owner publishes the output frequency and then either a + or a − indicator for the input signal frequency. The owner negotiates the choice of plus or minus through the repeater coordination group. In the 2-m band, the offset is typically 600 kHz, but it does not have to be. This makes **Answer B** the right choice. The other choices are wrong because they indicate a single offset direction, and/or the frequency offset is incorrect.

T2A02 What is the national calling frequency for FM simplex operations in the 2 meter band?

A. 146.520 MHz
B. 145.000 MHz
C. 432.100 MHz
D. 446.000 MHz

You should be able to eliminate Answers C and D immediately because those frequencies are in the 70-cm band. Next you need to memorize that the standard calling frequency in the 2-m band is 146.52 MHz, as in **Answer A**.

T2A03 What is a common repeater frequency offset in the 70 cm band?

A. Plus or minus 5 MHz
B. Plus or minus 600 kHz
C. Plus or minus 500 kHz
D. Plus or minus 1 MHz

This question is like the earlier one. Answers B has the values for the 2-m band standard, so be careful with it. The correct choice is the 5-MHz offset in **Answer A**. Answers C and D are to distract you.

T2A04 What is an appropriate way to call another station on a repeater if you know the other station's call sign?

A. Say "break, break" then say the station's call sign
B. Say the station's call sign then identify with your call sign
C. Say "CQ" three times then the other station's call sign
D. Wait for the station to call CQ then answer it

If you start listening to repeater traffic, you will notice that the users employ a quick protocol to initiate contact: the initiating station sends the call sign of the intended receiver, and then sends its own call sign, as in **Answer B**. The procedure does not use a "CQ," as one would use on the HF bands, so Answers C and D are incorrect. This is not a CB channel, so one does not use "break" making Answer A also incorrect.

T2A05 How should you respond to a station calling CQ?

A. Transmit "CQ" followed by the other station's call sign
B. Transmit your call sign followed by the other station's call sign
C. Transmit the other station's call sign followed by your call sign
D. Transmit a signal report followed by your call sign

Now we move from repeater operation to the regular contact on the bands. Here, the procedure is to respond to a "CQ" call with the other station's call followed by your call, as in **Answer C**. Be careful with Answer B because it is the opposite order from the correct response. One does not answer a "CQ" with another "CQ," so Answer A is incorrect. One does not answer a "CQ" with a signal report, so Answer D is also incorrect.

T2A06 Which of the following is required when making on-the-air test transmissions?

A. Identify the transmitting station
B. Conduct tests only between 10 p.m. and 6 a.m. local time
C. Notify the FCC of the test transmission
D. State the purpose of the test during the test procedure

The only requirement is to identify the transmitting station, as we saw in the previous chapter on the Part 97 rules, so **Answer A** is the correct choice. Answers B and C are to distract you. Answer D is being polite, so others do not respond, but Part 97 does not require it.

T2A07 What is meant by "repeater offset?"

A. The difference between a repeater's transmit frequency and its receive frequency
B. The repeater has a time delay to prevent interference
C. The repeater station identification is done on a separate frequency
D. The number of simultaneous transmit frequencies used by a repeater

The repeater offset is the frequency difference between the transmission and the receive frequencies, as in **Answer A**. The others are distractions to see if you understand repeaters.

T2A08 What is the meaning of the procedural signal "CQ"?
A. Call on the quarter hour
B. A new antenna is being tested (no station should answer)
C. Only the called station should transmit
D. Calling any station

The procedural sign (or *pro sign*) "CQ" by itself means you are inviting any station that can hear you to respond, so **Answer D** is the right choice. Be careful with Answer C because that is the correct response when you send their call sign followed by your call sign. Answers A and B are to distract you.

T2A09 What brief statement indicates that you are listening on a repeater and looking for a contact?
A. The words "Hello test" followed by your call sign
B. Your call sign
C. The repeater call sign followed by your call sign
D. The letters "QSY" followed by your call sign

Here we are back to the repeaters. To indicate you are listening, merely send your call sign, and see if someone responds, so **Answer B** is the correct choice. Answer D is incorrect because "QSY" indicates that you wish to change frequencies, and that makes no sense in this context. The others are to distract you.

T2A10 What is a band plan, beyond the privileges established by the FCC?
A. A voluntary guideline for using different modes or activities within an amateur band
B. A mandated list of operating schedules
C. A list of scheduled net frequencies
D. A plan devised by a club to indicate frequency band usage

To bring some operational order to the bands, the amateur community has developed band plans to share the bands. This plan is voluntary, and not one of the tables in Part 97, so **Answer A** is the right choice. You can find an example band plan at `http://www.arrl.org/band-plan`. The other choices are to distract you.

T2A11 What term describes an amateur station that is transmitting and receiving on the same frequency?
A. Full duplex
B. Diplex
C. Simplex
D. Multiplex

Be careful with the definitions here. A *full duplex* system allows transmission in both directions at the same time. While we want to establish two-way communications, we do not transmit and receive simultaneously. This makes Answer A incorrect. Since we do communicate in both directions, but only in one direction at a time, we have a *simplex* link, and **Answer C** is the correct choice. Multiplex communications, in Answer D, is a group of communications channels bundled together, so this is incorrect. When two transmitters or receivers share a common antenna, they are using the diplex operations of Answer B, so this is incorrect as well.

T2A12 Which of the following is a guideline to use when choosing an operating frequency for calling CQ?

A. Listen first to be sure that no one else is using the frequency
B. Ask if the frequency is in use
C. Make sure you are in your assigned band
D. All of these choices are correct

Each of the points in Answers A, B, and C are good amateur practice. This makes **Answer D** the best choice for this question.

2.3 T2B – VHF/UHF Operating Practices

2.3.1 Overview

The *VHF/UHF Operating Practices* question group in Subelement T2 concentrates on proper operating techniques in the VHF and UHF bands. This group covers topics such as

- SSB phone
- FM repeater
- Simplex
- Splits and shifts
- CTCSS
- DTMF
- Tone squelch
- Carrier squelch
- Phonetics
- Operational problem resolution
- Q signals

Some of the questions here build on the repeater questions from the previous group. There is a total of 14 questions in this group of which one will be selected for the exam.

2.3.2 Questions

T2B01 What is the most common use of the "reverse split" function of a VHF/UHF transceiver?

A. Reduce power output
B. Increase power output
C. Listen on a repeater's input frequency
D. Listen on a repeater's output frequency

Reverse split is not concerned with power output, so Answers A and B are incorrect. In this operation, the reverse split is listening on the repeater's input frequency making **Answer C** the correct choice. Answer D is the normal operational mode, not the reverse mode, so it is incorrect.

T2B02 What is the term used to describe the use of a sub-audible tone transmitted with normal voice audio to open the squelch of a receiver?

A. Carrier squelch
B. Tone burst
C. DTMF
D. CTCSS

Continuous Tone-Coded Squelch System (CTCSS) is the name for the tones used to open the squelch of a receiver like a repeater, so **Answer D** is the right choice. Dual Tone Multifrequency (DTMF) is the name for the tones like those found on your cell phone to dial or enter numbers, so Answer C is incorrect. Answers A and B are to distract you in this question.

T2B03 If a station is not strong enough to keep a repeater's receiver squelch open, which of the following might allow you to receive the station's signal?

A. Open the squelch on your radio
B. Listen on the repeater input frequency
C. Listen on the repeater output frequency
D. Increase your transmit power

Just because a repeater cannot properly receive a station, it does not mean you cannot communicate. You may be able to hear the station directly by listening on the repeater's input frequency, as in **Answer B**. Answer A is incorrect because the squelch on your radio does not match the repeater's squelch. The other station is not transmitting on the repeater's output frequency, so Answer C is incorrect. Raising your transmission power will not help the other station reach the repeater, so Answer D is not effective.

T2B04 Which of the following could be the reason you are unable to access a repeater whose output you can hear?

A. Improper transceiver offset
B. The repeater may require a proper CTCSS tone from your transceiver
C. The repeater may require a proper DCS tone from your transceiver
D. All of these choices are correct

This is a common problem for new operators or those using a repeater in an unfamiliar area. Each of the reasons given in Answers A, B, and C might cause the problem, so **Answer D** is the best choice here.

T2B05 What might be the problem if a repeater user says your transmissions are breaking up on voice peaks?

A. You have the incorrect offset
B. You need to talk louder
C. You are talking too loudly
D. Your transmit power is too high

With voice peaks, your signal may be "over deviating." That is, your FM signal bandwidth has grown as Figure 2.4 illustrates. This occurs when you are talking too loudly, as in **Answer C**. If you had the incorrect offset, as in Answer A, the repeater would not respond. If you were speaking softly, you would not have the break-up as in Answer B. If your power was too high, as in Answer D, you might have problems all the time and not just on peaks.

T2B06 What type of tones are used to control repeaters linked by the Internet Relay Linking Project (IRLP) protocol?

A. DTMF
B. CTCSS
C. EchoLink
D. Sub-audible

This is a characteristic of the Internet Radio Linking Project (IRLP) protocol that you will need to memorize. **Answer A** has the correct answer of DTMF tones. The other choices are incorrect for IRLP

T2B07 How can you join a digital repeater's "talk group"?

A. Register your radio with the local FCC office
B. Join the repeater owner's club
C. Program your radio with the group's ID or code
D. Sign your call after the courtesy tone

The Federal Communications Commission (FCC) will not resolve this issue for you, so Answer A is incorrect. Repeater's do not have "owner's clubs," so Answer B is incorrect. Repeaters do not have "courtesy tones," so Answer D is also incorrect.

Answer C correctly has the simple method of using the group's code.

T2B08 Which of the following applies when two stations transmitting on the same frequency interfere with each other?

A. Common courtesy should prevail, but no one has absolute right to an amateur frequency
B. Whoever has the strongest signal has priority on the frequency
C. Whoever has been on the frequency the longest has priority on the frequency
D. The station which has the weakest signal has priority on the frequency

We have two general principles from Part 97 to guide us here:

> Each station licensee and each control operator must cooperate in selecting transmitting channels and in making the most effective use of the amateur service frequencies. No frequency will be assigned for the exclusive use of any station.

and

> No amateur operator shall willfully or maliciously interfere with or cause interference to any radio communication or signal.

With this, **Answer A** gives the best solution, and operators need to have a courteous approach to operating and cooperatively share the bands. The other choices do not follow best amateur practice, so they are incorrect.

T2B09 What is a "talk group" on a DMR digital repeater?

A. A group of operators sharing common interests
B. A way for groups of users to share a channel at different times without being heard by other users on the channel
C. A protocol that increases the signal-to-noise ratio when multiple repeaters are linked together
D. A net that meets at a particular time

The Digital Mobile Radio (DMR) protocol permits establishing user groups to channel share without being heard by other users as in **Answer B**. Answers A and D do not need DMR for amateurs to do those activities, so these are incorrect. Answer C is a distraction.

T2B10 Which Q signal indicates that you are receiving interference from other stations?

A. QRM
B. QRN
C. QTH
D. QSB

Going back to Table 2.1, we see that QRM, as in **Answer A**, is correct. QRN indicates static, QTH is for the location, and QSB is for fading.

T2B11 Which Q signal indicates that you are changing frequency?
A. QRU
B. QSY
C. QSL
D. QRZ

Again, we go back to Table 2.1, and see that QSY, as in **Answer B**, is correct. QRU is asking for messages, QSL acknowledges receipt, and QRZ is asking who is calling.

T2B12 Why are simplex channels designated in the VHF/UHF band plans?
A. So that stations within mutual communications range can communicate without tying up a repeater
B. For contest operation
C. For working DX only
D. So that stations with simple transmitters can access the repeater without automated offset

If you think about it, it makes sense to communicate without a repeater if you don't really need it, so **Answer A** is the right choice. Operators do not use repeaters in contests or for DX, so Answers B and C are incorrect. Answer D does not correspond to amateur radio hardware.

T2B13 Where may SSB phone be used in amateur bands above 50 MHz?
A. It is permitted only by holders of a General Class or higher license
B. It is permitted only on repeaters
C. It is permitted in at least some portion of all the amateur bands above 50 MHz
D. It is permitted only on when power is limited to no more than 100 watts

Answer A is incorrect because Technician Class operators can use phone transmissions. If you look over Part 97 in detail, you will see that Part 97 allocates some portion of the bands above 50 MHz for phone, so **Answer C** is correct. Answer B is also incorrect because operators can use phone without repeaters. Answer D does not match Part 97, so it is incorrect.

T2B14 Which of the following describes a linked repeater network?
A. A network of repeaters where signals received by one repeater are repeated by all the repeaters
B. A repeater with more than one receiver
C. Multiple repeaters with the same owner
D. A system of repeaters linked by APRS

As the name implies, a linked network has the repeaters linked together so that all repeaters in the network can repeat the signals. This makes **Answer A** the correct choice. The other choices are distractors to see if you understand repeaters.

2.4 T2C – Public Service

2.4.1 Overview

The *Public Service* question group in Subelement T2 concentrates on Amateur Radio Service support for emergency communications. The *Public Service* question covers topics such as

- Emergency and non-emergency operations
- Applicability of FCC rules
- RACES and ARES
- Net and traffic procedures
- Operating restrictions during emergencies

These topics will refer to the FCC Part 97 and operating practice. There is a total of 12 questions in this section of which one will be selected for the exam.

2.4.2 Questions

T2C01 [97.103(a)] When do the FCC rules NOT apply to the operation of an amateur station?

A. When operating a RACES station
B. When operating under special FEMA rules
C. When operating under special ARES rules
D. Never, FCC rules always apply

Part 97 is very clear that "The station licensee is responsible for the proper operation of the station in accordance with the FCC Rules." This means that the FCC rules always apply, as in **Answer D**. Everything else is not in compliance with Part 97.

T2C02 What is meant by the term "NCS" used in net operation?

A. Nominal Control System
B. Net Control Station
C. National Communications Standard
D. Normal Communications Syntax

Every net has a Net Control Station (NCS), which makes **Answer B** the correct choice. The other options are to distract you.

T2C03 What should be done when using voice modes to ensure that voice messages containing unusual words are received correctly?

A. Send the words by voice and Morse code
B. Speak very loudly into the microphone
C. Spell the words using a standard phonetic alphabet
D. All of these choices are correct

In this case, using the phonetic alphabet, as in **Answer C**, follows that good operating procedure. Answers A and B do not provide good results, so they are incorrect. This also makes Answer D incorrect.

T2C04 What do RACES and ARES have in common?

A. They represent the two largest ham clubs in the United States
B. Both organizations broadcast road and weather information
C. Neither may handle emergency traffic supporting public service agencies
D. Both organizations may provide communications during emergencies

We saw Radio Amateur Civil Emergency Service (RACES) in subelement T1. Amateur Radio Emergency Service (ARES) is another emergency communications organization. This makes **Answer D** the right choice.

T2C05 What does the term "traffic" refer to in net operation?

A. Formal messages exchanged by net stations
B. The number of stations checking in and out of a net
C. Operation by mobile or portable stations
D. Requests to activate the net by a served agency

In nets, the "traffic " is the message traffic the operators are exchanging as in **Answer A**. The other options are incorrect statements to distract you.

T2C06 Which of the following is an accepted practice to get the immediate attention of a net control station when reporting an emergency?

A. Repeat "SOS" three times followed by the call sign of the reporting station
B. Press the push-to-talk button three times
C. Begin your transmission by saying "Priority" or "Emergency" followed by your call sign
D. Play a pre-recorded emergency alert tone followed by your call sign

Answers A, B, and D all represent bad operating practice, so they are incorrect. **Answer C** gives the correct procedure.

T2C07 Which of the following is an accepted practice for an amateur operator who has checked into a net?

A. Provided that the frequency is quiet, announce the station call sign and location every 5 minutes
B. Move 5 kHz away from the net's frequency and use high power to ask other hams to keep clear of the net frequency
C. Remain on frequency without transmitting until asked to do so by the net control station
D. All of the choices are correct

Answers A and B represent poor operating practice, so they are incorrect. This also makes Answer D an incorrect choice. **Answer C** shows you the correct practice to maintain proper net etiquette (or netiquette).

T2C08 Which of the following is a characteristic of good traffic handling?

A. Passing messages exactly as received
B. Making decisions as to whether messages are worthy of relay or delivery
C. Ensuring that any newsworthy messages are relayed to the news media
D. All of these choices are correct

Answer B defeats the purpose, so this is incorrect. Answer C could be a violation of Part 97, so this is not a good choice either. These make Answer D incorrect. **Answer A** shows you the proper procedure.

T2C09 Are amateur station control operators ever permitted to operate outside the frequency privileges of their license class?

A. No
B. Yes, but only when part of a FEMA emergency plan
C. Yes, but only when part of a RACES emergency plan
D. Yes, but only if necessary in situations involving the immediate safety of human life or protection of property

Part 97 tells us that

> No provision of these rules prevents the use by an amateur station of any means of radiocommunication at its disposal to provide essential communication needs in connection with the immediate safety of human life and immediate protection of property when normal communication systems are not available.

Answer D is consistent with this section of Part 97. The other choices are not.

T2C10 What information is contained in the preamble of a formal traffic message?

A. The email address of the originating station
B. The address of the intended recipient
C. The telephone number of the addressee
D. The information needed to track the message

In many messaging protocols, the preamble is a message header with information used for routing and accounting; it is not part of the message information section. **Answer D** matches this concept, so it is the right choice. Answers A, B, and C are not parts of traffic messages, so they are incorrect choices.

T2C11 What is meant by the term "check" in reference to a formal traffic message?

A. The check is a count of the number of words or word equivalents in the text portion of the message
B. The check is the value of a money order attached to the message
C. The check is a list of stations that have relayed the message
D. The check is a box on the message form that tells you the message was received and/or relayed

In this sense, the "check" is an error detection means. That is, operators can use the check to tell if the message has lost information, so **Answer A** is the correct choice. The other answers are to distract you.

T2C12 What is the Amateur Radio Emergency Service (ARES)?

A. Licensed amateurs who have voluntarily registered their qualifications and equipment for communications duty in the public service
B. Licensed amateurs who are members of the military and who voluntarily agreed to provide message handling services in the case of an emergency
C. A training program that provides licensing courses for those interested in obtaining an amateur license to use during emergencies
D. A training program that certifies amateur operators for membership in the Radio Amateur Civil Emergency Service

We saw ARES in an earlier question, so you should be able to spot **Answer A** as the right choice. The other choices are useful groups, but not ARES.

Chapter 3

T3 – RADIO WAVE CHARACTERISTICS

3.1 Introduction

So far, we have been dealing with rules and amateur radio practice. In this chapter, the questions aimed at testing your understanding of the definition of radio waves, and how they travel in free space and in our environment. While the questions are mostly concerned with how the radio signals affect our amateur operations, this knowledge will also help you understand other applications such as commercial radio and television broadcasts, and the Wi-Fi routers in our homes and offices. This *Radio Wave Characteristics* subelement has the following question groups:

A. Radio wave characteristics
B. Radio and electromagnetic wave properties
C. Propagation modes

This will generate three questions on the Technician examination.

3.2 Radio Engineering Concepts

The questions in this subelement cover our understanding of what makes up a radio wave and how those waves travel. The atmosphere and objects in our environment interact with the radio waves and cause both useful and bad effects, depending upon the geometry and the wavelength, or frequency, of the radio wave. We start with a description of the radio wave and how it travels in our environment. Then we proceed to understanding the atmosphere and characteristics of the radio waves in the different bands.

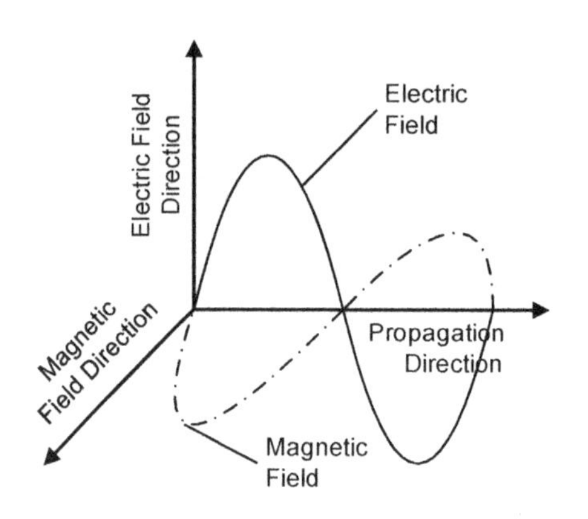

Figure 3.1: The electromagnetic radiation wave. The electric and magnetic fields are perpendicular to each other and the direction of travel.

Electromagnetic Waves Radio waves are a type of electromagnetic radiation just as visible light is. The main difference is that the wavelength of visible light is on the order of the size of molecules, while the wavelength of radio waves used by amateurs typically ranges from the size of your hand to the length of a football field. The wave has two components: an electrical field and a magnetic field. They are aligned at 90° angles to each other and to the direction of travel. Figure 3.1 shows these fields.

As we saw in Chapter 2, radio waves move at the speed of light or 299 792 458 m/s. The defining relationship between the radio wave's wavelength, λ, frequency, f, and the speed of light, c, is $c = \lambda f$.

Polarization The orientations of the electric and magnetic fields are important because antennas are sensitive to those characteristics. Also, parts of the atmosphere will interact with the fields and modify these orientations. We track the orientation through the concept of the radio wave's *polarization*. If the transmitting antenna and the receiving antenna do not have their natural polarizations aligned, then we can have substantial signal loss that degrades communications.

The electric field orientation is the standard for tracking the polarization. If the field maintains the same orientation plane as it travels, then the wave has *linear polarization*. If the linear polarization is vertical to the Earth's surface then we have *vertical polarization*, while if it is parallel to the Earth's surface, it is *horizontal polarization*.

The electric filed is not required to maintain a fixed plane. We can imagine that the electric field in Figure 3.1 is spinning around the direction of propagation. If that is the case, then we have *circular polarization*. The spin can go in either direction giving either *right hand* or *left hand* circular polarization.

Finally, the electromagnetic wave may be a mix of circular and linear polarization, which gives *elliptical polarization*.

Atmospheric Structure Radio waves propagate through the atmosphere. The structure of the atmosphere affects the quality of the signals and the ways the signals move from the transmitter to the receiver. Figure 3.2 presents a diagram of the atmosphere's layers. Amateur radio operators consider two major atmospheric regions for supporting their communications: the troposphere and the ionosphere. Figure 3.2 shows the troposphere is the region where we live and where the weather

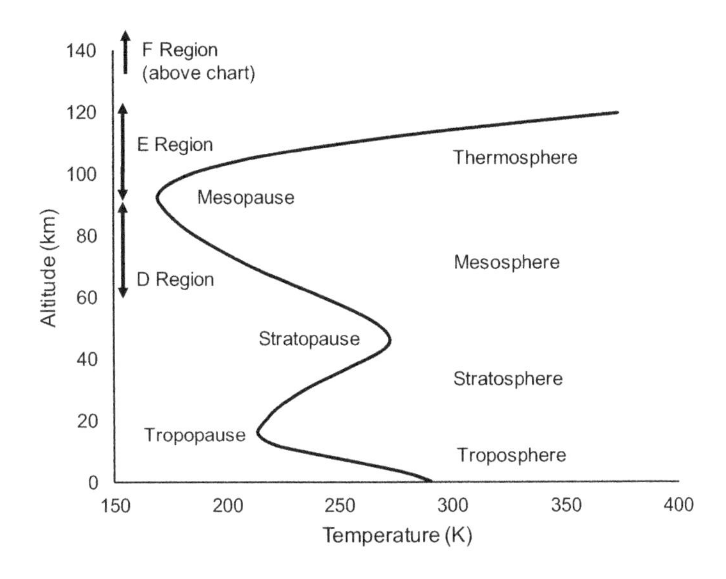

Figure 3.2: Layers in the Earth's atmosphere.

occurs. The ionosphere is above the troposphere and is the region where atmospheric gases become ionized. Scientists divide it into layers or regions where the electron density is relatively high. The ionosphere's layers that are of concern for amateur radio operations are

- *D* region at approximately 80 km; it is important during the day and fades away at night
- *E* region at approximately 110 km
- *F* region which forms two areas during the day, *F1* at 200 km and *F2* at 300 km, and it fades into a single region at night

The ionosphere influences radio propagation by the disturbances that are functions of solar activity, time of day, frequency, and the angle at which the radio waves enter.

Radio Propagation Amateur radio operators need to understand how the radio waves travel through the atmosphere and permit communications. If you are new to radio activities, you may think that the Radio Frequency (RF) radiation just travels along a straight line from point to point. However, in practice, the situation is more complicated, as Figure 3.3 illustrates. Radio waves may travel between two antennas via one or more of four propagation modes

Sky waves — Radio waves that reflect off the Earth's atmosphere

Line of sight — Radio waves that travel along a direct path between the two antennas

Reflected waves — Radio waves that reflect off the surface of the Earth or structures in the environment

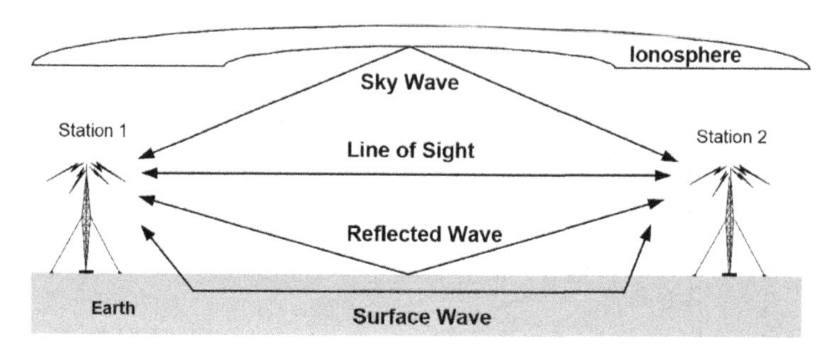

Figure 3.3: Radio propagation modes through our environment.

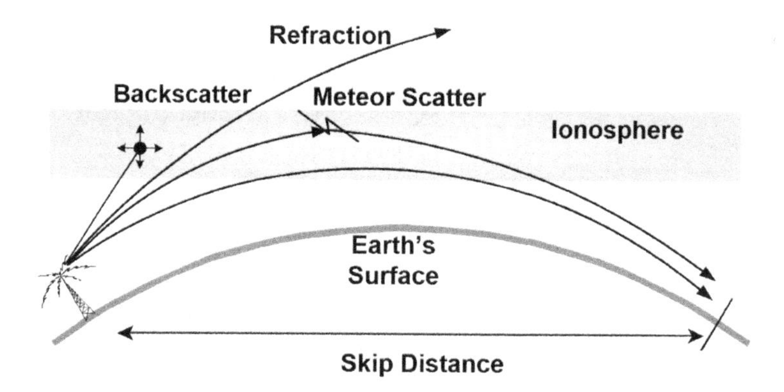

Figure 3.4: Various forms of sky wave propagation.

Surface waves — Radio waves that travel along the Earth's surface
Radio operators use each of these modes at various times.

Sky Wave Propagation Amateur radio operators depend on sky wave propagation to support communications around the world. There are several forms of sky wave propagation that are of interest to us, as Figure 3.4 illustrates:

Ionospheric Reflection — Radio waves that reflect off a layer of the Earth's ionosphere

Sporadic E Reflection — Radio waves that scatter off an ionization pocket region in the E layer of the ionosphere

Meteor Scatter — Radio waves that reflect off the ionized trails of meteors passing through the ionosphere

Tropospheric Scatter — Radio waves that use scattering of the RF signal by the Earth's troposphere instead of the ionosphere

Atmospheric Ducting — Radio waves that travel along a path between inversion layers in the Earth's troposphere

Ionospheric reflection is the most common and the most useful, but the others are great to use when conditions permit.

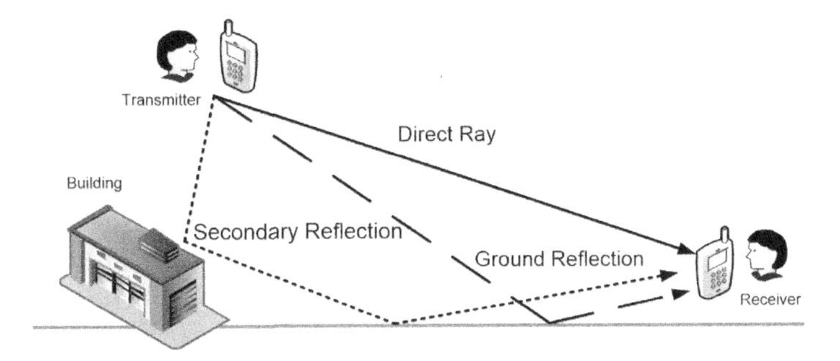

Figure 3.5: Diagram for potential multipath propagation paths between a transmitter and receiver.

Refraction and Diffraction There are two sky wave effects that radio operators encounter on a regular basis across many of the bands: refraction and diffraction. *Refraction* is one of the modes that Figure 3.4 illustrates. It is the changing arced path of the radio waves. When an electromagnetic wave moves through any medium (air, water, etc.), changes in the medium's density affects its path. The Earth's atmosphere changes its density with height above sea level and with weather, so the RF radiation also changes its path with this changing density. With *diffraction*, the path of the electromagnetic wave interacts with a relatively sharp edge such as a building or a hill. This edge causes the radio path to scatter around it, which permits propagation into a region where you might think the signal is blocked. In this case, the signal may not be strong, but it will be there.

Multipath Propagation Not every radio transmission exclusively follows the direct path between the transmitter and the receiver. Multipath propagation is a combination of line-of-sight propagation with reflected-wave propagation. Because the transmitter's antenna spreads the radiation out over a wider area, that radiation interacts with buildings, the ground, and land features. Some of the radiation reflects off these items and into the receiving antenna along with the direct radiation, as Figure 3.5 shows. Because the distances along the various paths are different, the radio waves do not necessarily add together in a favorable manner in the receiver. This can cause fades on the signal. Increasing power is not usually the best answer — moving locations can be more effective.

Band Characteristics Table 3.1 shows the dominant propagation characteristics in the different bands. Lower frequencies give the operator longer propagation distances. However, noise, amplitude fluctuations, and other effects affect links over these long distances. The high-frequency carriers force the operator to use only line-of-sight communications. If the operator also needs long distances with the high frequencies, then the operator needs some form of relay network.

Table 3.1: Propagation Characteristics of Radio Bands

Band	Frequency	Characteristic
High Frequency (HF)	3 to 30 MHz	Atmospheric noise, ionospheric reflections, long-distance links; affected by solar flux
Very High Frequency (VHF)	30 to 300 MHz	Some ionospheric reflections, sporadic E, and meteor scatter; mostly line-of-sight propagation
Ultra High Frequency (UHF)	300 to 3000 MHz	Mostly line-of-sight propagation
Super High Frequency (SHF)	3 to 30 GHz	Line-of-sight propagation; atmospheric absorption at higher frequencies
Extremely High Frequency (EHF)	30 to 300 GHz	Line-of-sight propagation; subject to atmospheric absorption

Band Designations Several questions in this sub-element refer to the official band designations from the International Telecommunication Union (ITU) *Radio Regulations*. Table 3.2 lists these official designations.

3.3 T3A – Radio Wave Characteristics

3.3.1 Overview

The **Radio Wave Characteristics** question group in Subelement T3 introduces you the general radio wave characteristics that affect transmitting signals for amateur radio operations. The *Radio Wave Characteristics* group covers topics such as

- How a radio signal travels
- Fading
- Multipath
- Polarization
- Wavelength vs absorption
- Antenna orientation

There is a total of 13 questions in this group of which one will be selected for the exam.

Table 3.2: Standard Radio Band Designations

Band Number	Designation	Frequency Range
3	Extremely Low Frequency (ELF)	<3 kHz
4	Very Low Frequency (VLF)	3 to 30 kHz
5	Low Frequency (LF)	30 to 300 kHz
6	Medium Frequency (MF)	300 to 3000 kHz
7	High Frequency (HF)	3 to 30 MHz
8	Very High Frequency (VHF)	30 to 300 MHz
9	Ultra High Frequency (UHF)	300 to 3000 MHz
10	Super High Frequency (SHF)	3 to 30 GHz
11	Extremely High Frequency (EHF)	30 to 300 GHz
12	Sub-millimeter	300 to 3000 GHz

3.3.2 Questions

T3A01 What should you do if another operator reports that your station's 2 meter signals were strong just a moment ago, but now they are weak or distorted?

A. Change the batteries in your radio to a different type
B. Turn on the CTCSS tone
C. Ask the other operator to adjust his squelch control
D. Try moving a few feet or changing the direction of your antenna if possible, as reflections may be causing multi-path distortion

Answer A may sound like the right choice, but we are discussing radio propagation in this section, so the batteries do not count. Radios use the Continuous Tone-Coded Squelch System (CTCSS) tones in Answer B to keep users without the proper access tone from being on the channel. They will not affect the signal's quality, so this is incorrect. The squelch control on Answer C cuts the audio on and off, but does not cause this kind of distortion. This type of distortion characterizes the multi-path of **Answer D** making this is the correct choice.

T3A02 Why might the range of VHF and UHF signals be greater in the winter?

A. Less ionospheric absorption
B. Less absorption by vegetation
C. Less solar activity
D. Less tropospheric absorption

If you have satellite radio in your car, you may already know the answer to this question. Vegetation, like trees, can absorb the VHF and UHF RF signals. In the winter, there is less vegetation to absorb the signals than in the summer. This makes **Answer B** correct. The Earth's seasons do not affect the sun, so Answer C is incorrect. VHF and UHF signals generally pass through the atmosphere, so Answers A and D are incorrect.

T3A03 What antenna polarization is normally used for long-distance weak-signal CW and SSB contacts using the VHF and UHF bands?

A. Right-hand circular
B. Left-hand circular
C. Horizontal
D. Vertical

Operators often use circular polarizations in satellite communications, but not on ground-based long-distance communications, so Answers A and B are incorrect choices. You need to remember that horizontal polarization works best in this application, so **Answer C** is the right choice. The vertical polarization in Answer D is not as good of a choice.

T3A04 What can happen if the antennas at opposite ends of a VHF or UHF line of sight radio link are not using the same polarization?

A. The modulation sidebands might become inverted
B. Signals could be significantly weaker
C. Signals have an echo effect on voices
D. Nothing significant will happen

The technical name for this is a "polarization mismatch" and it causes a loss in signal strength, so **Answer B** is the right choice. Answers A and C indicate some form of distortion, so they are incorrect. Since there is signal loss, Answer D is also incorrect.

T3A05 When using a directional antenna, how might your station be able to access a distant repeater if buildings or obstructions are blocking the direct line of sight path?

A. Change from vertical to horizontal polarization
B. Try to find a path that reflects signals to the repeater
C. Try the long path
D. Increase the antenna SWR

When one is using a directional antenna, the antenna confines the RF radiation to a relatively narrow beamwidth — think of a flashlight beam. In this question, the implication is that we cannot access the repeater directly, so how do we do it indirectly? Try bouncing the RF signal off reflective object (certain buildings are very good for this in a city), as **Answer B** suggests. Changing the polarization will not help with obstructions, so Answer A is incorrect. Since repeaters generally use high RF frequencies, taking the long path around the Earth like one can do at HF frequencies is not possible, so Answer C is not a good choice. Increasing the Standing Wave Ratio (SWR) will increase the mismatch between your transmitter and antenna, and only increase system loss, if not also potentially damaging your equipment, so Answer D is a very bad choice.

T3A06 What term is commonly used to describe the rapid fluttering sound sometimes heard from mobile stations that are moving while transmitting?

A. Flip-flopping
B. Picket fencing
C. Frequency shifting
D. Pulsing

This is a term of practice that you will need to remember. This is the classical description of a picket fence fade, as in **Answer B**, so that is the right choice. Be careful with Answer D because it sounds plausible, but that is not what operators call this audio effect, so it is incorrect. The frequency shifting of Answer C will disrupt the signal, but not in this way, so it is incorrect. The flip-flopping in Answer A is to distract you.

T3A07 What type of wave carries radio signals between transmitting and receiving stations?

A. Electromagnetic
B. Electrostatic
C. Surface acoustic
D. Magnetostrictive

This should be an easy one to pick out: RF signals are electromagnetic waves, as in **Answer A**. Be careful with Answer B because electrostatics set up electric fields and sound similar, but they do not produce waves, so the choice is incorrect. The surface acoustic waves in Answer C are sound waves, so this is an incorrect choice. The magnetostriction of Answer D is a property of compounds like iron, but is not a cause of RF waves.

T3A08 Which of the following is a likely cause of irregular fading of signals received by ionospheric reflection?

A. Frequency shift due to Faraday rotation
B. Interference from thunderstorms
C. Random combining of signals arriving via different paths
D. Intermodulation distortion

We have discussed ionospheric reflections, and sometimes those reflections are smooth like a mirror, while sometimes they are more like a broken mirror. This is an example of the latter, so there will be a random signal characteristic caused by this variable reflection in the ionosphere. Therefore, **Answer C** is the correct choice. The Faraday rotation in Answer A is a polarization shift caused by the Earth's magnetic field that does not cause random fades, so Answer A is incorrect. The thunderstorms of Answer B do not reach the ionosphere, so this is also incorrect. The intermodulation distortion can affect signals, but your equipment causes it, and not the ionosphere. Answer D is also incorrect.

T3A09 Which of the following results from the fact that skip signals refracted from the ionosphere are elliptically polarized?

A. Digital modes are unusable
B. Either vertically or horizontally polarized antennas may be used for transmission or reception
C. FM voice is unusable
D. Both the transmitting and receiving antennas must be of the same polarization

Elliptical polarization causes the polarization state of the electromagnetic wave to be between that of purely vertical and purely horizontal polarization. This means that antennas using either polarization will work, as in **Answer B**. If Answer B is correct, then Answer D must be incorrect. Saying the voice and data modes will not work is not correct engineering, so Answers A and C are also incorrect.

T3A10 What may occur if data signals propagate over multiple paths?

A. Transmission rates can be increased by a factor equal to the number of separate paths observed
B. Transmission rates must be decreased by a factor equal to the number of separate paths observed
C. No significant changes will occur if the signals are transmitting using FM
D. Error rates are likely to increase

Multiple paths make for multiple copies of the signal. With digital modes, this leads to an increase in errors as the multiple signals add together in the receiver, and the signal becomes corrupted. In this case, **Answer D** is correct, which also makes Answer C incorrect. Answers A and B are silly engineering distractors.

T3A11 Which part of the atmosphere enables the propagation of radio signals around the world?

A. The stratosphere
B. The troposphere
C. The ionosphere
D. The magnetosphere

Ions in the ionosphere cause the reflections that produce the skips to enable radio propagation around the world. This makes **Answer C** the right choice. The other choices are to distract you.

T3A12 How might fog and light rain affect radio range on the 10 meter and 6 meter bands?

A. Fog and rain absorb these wavelength bands
B. Fog and light rain will have little effect on these bands
C. Fog and rain will deflect these signals
D. Fog and rain will increase radio range

Rain drops will absorb the RF signal, especially for frequencies above 20 GHz. Both 10-m and 6-m RF signals are well below this frequency, so they will have little absorption as in **Answer B**. Answer A would be true at the higher frequencies. Answers C and D are to distract you.

T3A13 What weather condition would decrease range at microwave frequencies?
A. High winds
B. Low barometric pressure
C. Precipitation
D. Colder temperatures

Based on the previous question, did you pick precipitation from **Answer C** as the right choice? Wind and temperature will not affect RF signal absorption making Answers A and D incorrect. Low barometric pressure may be associated with a rain storm, but the pressure by itself will not produce absorption, so Answer B is not a good choice.

3.4 T3B – Radio and Electromagnetic Wave Properties

3.4.1 Overview

The *Radio and Electromagnetic Wave Properties* question group in Subelement T3 introduces you to the physical characteristics of radio waves. The *Radio and Electromagnetic Wave Properties* group covers topics such as

- The electromagnetic spectrum
- Wavelength vs. frequency
- Nature and velocity of electromagnetic waves
- Definition of UHF, VHF, HF bands
- Calculating wavelength

There is a total of 11 questions in this group of which one will be selected for the exam.

3.4.2 Questions

T3B01 What is the name for the distance a radio wave travels during one complete cycle?
A. Wave speed
B. Waveform
C. Wavelength
D. Wave spread

This one should be easy to figure out: distance is length, so the question is asking about the wavelength of the RF wave. This makes **Answer C** the right answer. This is analogous to the "Period" in Figure 2.2. Later questions will address the speed, so

Answer A is incorrect here. Waveform is the wave's shape not length, so Answer B is incorrect. The wave spread of Answer D is a distortion measure, so it is also incorrect here.

T3B02 What property of a radio wave is used to describe its polarization?
A. The orientation of the electric field
B. The orientation of the magnetic field
C. The ratio of the energy in the magnetic field to the energy in the electric field
D. The ratio of the velocity to the wavelength

The RF wave's electric field defines the type of polarization in the wave, so **Answer A** is the right answer. The wave's magnetic field does not define the polarization, so Answer B is incorrect. Answers C and D are silly engineering-sounding distractors.

T3B03 What are the two components of a radio wave?
A. AC and DC
B. Voltage and current
C. Electric and magnetic fields
D. Ionizing and non-ionizing radiation

As we said earlier, the radio wave has two components: the electric field and the magnetic field. This makes **Answer C** correct. Answers A and B will be important when we get to circuits, but they are incorrect here. Radio waves are examples of non-ionizing radiation, so Answer D is to distract you.

T3B04 How fast does a radio wave travel through free space?
A. At the speed of light
B. At the speed of sound
C. Its speed is inversely proportional to its wavelength
D. Its speed increases as the frequency increases

This should be an obvious choice: radio waves travel at the speed of light, as in **Answer A**. Radio waves are not sound waves, so Answer B is incorrect. Answers C and D are silly physics-sounding distractors.

T3B05 How does the wavelength of a radio wave relate to its frequency?
A. The wavelength gets longer as the frequency increases
B. The wavelength gets shorter as the frequency increases
C. There is no relationship between wavelength and frequency
D. The wavelength depends on the bandwidth of the signal

Referring to the beginning of the chapter, the relationship between the radio wave's wavelength, λ, frequency, f, and the speed of light, c, is $c = \lambda f$. From this, we can see that to maintain the constant speed c when the wavelength shortens, the frequency must increase, as in **Answer B**. Be careful with Answer A because it is similar, but it does not carry the correct mathematical relationship. Answers C and D represent bad physics distractions.

T3B06 What is the formula for converting frequency to approximate wavelength in meters?

A. Wavelength in meters equals frequency in hertz multiplied by 300
B. Wavelength in meters equals frequency in hertz divided by 300
C. Wavelength in meters equals frequency in megahertz divided by 300
D. Wavelength in meters equals 300 divided by frequency in megahertz

Here we need to do some **very** rough approximations. The speed of light is around 300×10^6 m/s (or 299 792 458 m/s exactly). We rewrite the wavelength-frequency equation as $\lambda = c/f$. If we express the frequency in MHz, then the equation becomes $\lambda = 300/f(MHz)$, as in **Answer D**. Be careful because all the distraction answers look similar, but represent math mistakes.

T3B07 What property of radio waves is often used to identify the different frequency bands?

A. The approximate wavelength
B. The magnetic intensity of waves
C. The time it takes for waves to travel one mile
D. The voltage standing wave ratio of waves

Operators use both the wavelength and the frequency of the radio wave to identify which band the RF signal is in. Of the choices given here, **Answer A**, the wavelength, is the right choice. The SWR in Answer D is important in characterizing your system's performance, but it does not determine the operating band, so it is incorrect. Choices B and C are distractors.

T3B08 What are the frequency limits of the VHF spectrum?

A. 30 to 300 kHz
B. 30 to 300 MHz
C. 300 to 3000 kHz
D. 300 to 3000 MHz

Here, you may wish to refer to Table 3.2 for the band designations. From the table, we see that **Answer B** is the right choice for VHF. Answer A is the LF band. Answer C is the MF band. Answer D is the UHF band.

T3B09 What are the frequency limits of the UHF spectrum?

A. 30 to 300 kHz
B. 30 to 300 MHz
C. 300 to 3000 kHz
D. 300 to 3000 MHz

As we saw in Table 3.2, the UHF band runs 300 to 3000 MHz, so **Answer D** is the right choice. Answer A is the LF band. Answer B is the VHF band. Answer C is the MF band.

T3B10 What frequency range is referred to as HF?
- A. 300 to 3000 MHz
- B. 30 to 300 MHz
- C. 3 to 30 MHz
- D. 300 to 3000 kHz

This is a new one. The HF band runs 3 to 30 MHz, so **Answer C** is the right choice. Answer A is the UHF band. Answer B is the VHF band. Answer C is the MF band.

T3B11 What is the approximate velocity of a radio wave as it travels through free space?
- A. 150,000 kilometers per second
- B. 300,000,000 meters per second
- C. 300,000 miles per hour
- D. 150,000 miles per hour

From our earlier discussion, we should be able to spot the 300,000,000 meters per second in **Answer B** as the right choice. The other distraction choices have the wrong numbers or the wrong units.

3.5 T3C - Propagation Modes

3.5.1 Overview

The *Propagation Modes* question group in Subelement T3 introduces you to some of the various propagation modes we encounter in amateur radio operations. The *Propagation Modes* group covers topics such as
- Line of sight
- Sporadic E
- Meteor and auroral scatter and reflections
- Tropospheric ducting
- F layer skip
- Radio horizon

There is a total of 11 questions in this group of which one will be selected for the exam.

3.5.2 Questions

T3C01 Why are direct (not via a repeater) UHF signals rarely heard from stations outside your local coverage area?

A. They are too weak to go very far
B. FCC regulations prohibit them from going more than 50 miles
C. UHF signals are usually not reflected by the ionosphere
D. UHF signals are absorbed by the ionospheric D layer

Radio waves above a certain frequency generally pass through the atmosphere, and are not reflected. UHF frequencies are usually in this class of signals, so **Answer C** is the right choice. Answer A is untrue from a physics point of view. While the Federal Communications Commission (FCC) is very important, it cannot dictate the laws of physics, so Answer B is a funny distractor. Answer D is incorrect because the UHF signals pass through the atmosphere without the D layer absorbing them.

T3C02 Which of the following is an advantage of HF vs VHF and higher frequencies?

A. Signals are being reflected from outer space
B. Signals are arriving by sub-surface ducting
C. Signals are being reflected by lightning storms in your area
D. Signals are being refracted from a sporadic E layer

Generally, VHF and higher-frequency signals will pass through the atmosphere without the ionosphere reflecting them. The HF signals will be reflected by elements of the Earth's atmosphere. HF signals can be reflected by lightening while VHF is not, making **Answer C** the correct choice. Outer space does not reflect any RF band, so Answer A is a silly distractor. Neither band uses sub-surface ducting, so Answer B is also a distractor. VHF signals can be reflected by sporadic E, but not HF, so Answer D is the opposite of the question.

T3C03 What is a characteristic of VHF signals received via auroral reflection?

A. Signals from distances of 10,000 or more miles are common
B. The signals exhibit rapid fluctuations of strength and often sound distorted
C. These types of signals occur only during winter nighttime hours
D. These types of signals are generally strongest when your antenna is aimed west

If you have watched an aurora, you will remember that it is not static, but is constantly changing. Since it represents a charged phenomenon in the atmosphere, you should expect that an auroral reflection will have variability like **Answer B** describes, so that is the right choice. Auroras do not happen only in the winter, so Answer C is incorrect. Answers A and D are incorrect engineering statements.

T3C04 Which of the following propagation types is most commonly associated with occasional strong over-the-horizon signals on the 10, 6, and 2 meter bands?
A. Backscatter
B. Sporadic E
C. D layer absorption
D. Gray-line propagation

Normally, frequencies above 10 meters, especially 6 meters and 2 meters, have difficulty with strong over the horizon propagation. That is, unless the ionosphere is especially favorable, which occurs when there is a "Sporadic E" opening, as in **Answer B**. The other three choices are legitimate propagation effects, but none of them specifically give the conditions mentioned in the question. All of these are for the lower frequency bands below the VHF bands (the 10-meter band is just below the VHF band).

T3C05 Which of the following effects might cause radio signals to be heard despite obstructions between the transmitting and receiving stations?
A. Knife-edge diffraction
B. Faraday rotation
C. Quantum tunneling
D. Doppler shift

Knife-edge diffraction is when a radio wave hits an obstruction and the obstruction scatters it forward along the path. This permits reception beyond an obstruction, so **Answer A** is the right choice. The Faraday rotation of Answer B is a polarization effect, and the Doppler shift of Answer D is a frequency shift. Neither will give further propagation, so they are incorrect. The quantum tunneling of Answer C is a semiconductor device phenomenon, so this is also incorrect.

T3C06 What mode is responsible for allowing over-the-horizon VHF and UHF communications to ranges of approximately 300 miles on a regular basis?
A. Tropospheric scatter
B. D layer refraction
C. F2 layer refraction
D. Faraday rotation

This is like the Sporadic E of the earlier question, but this time the range is shorter, so we are dealing with a lower atmospheric layer (see Figure 3.2). Here we are seeing tropospheric scatter, so **Answer A** is the right choice. The D and F layers too high to give a reflection over the stated distance, so Answers B and C are incorrect. The Faraday rotation in Answer D is a distraction again.

T3C07 What band is best suited for communicating via meteor scatter?

A. 10 meters
B. 6 meters
C. 2 meters
D. 70 cm

This is one you will learn best by operating experience. Meteors ionize the upper atmosphere as they pass. The 6-meter band is the most common place to find operators using these meteor paths for sending communications, so **Answer B** is the right choice. The other choices are not as useful in this mode, so they are not correct choices.

T3C08 What causes tropospheric ducting?

A. Discharges of lightning during electrical storms
B. Sunspots and solar flares
C. Updrafts from hurricanes and tornadoes
D. Temperature inversions in the atmosphere

With tropospheric ducting, the atmosphere forms two layers that the radio wave bounces between until it finally exits. It is like a guided channel for the radio wave. An inversion layer is the usual cause of this ducting, which makes **Answer D** the correct answer. Answers A and B can affect radio wave propagation, but not help in creating a tropospheric duct, so they are incorrect choices. Answer C is to distract you.

T3C09 What is generally the best time for long-distance 10 meter band propagation via the F layer?

A. From dawn to shortly after sunset during periods of high sunspot activity
B. From shortly after sunset to dawn during periods of high sunspot activity
C. From dawn to shortly after sunset during periods of low sunspot activity
D. From shortly after sunset to dawn during periods of low sunspot activity

The F layer is stimulated by sunspot activity and it tends to collapse at night. We want to choose an answer that covers daytime and favorable sunspot activity, as **Answer A** does. Be careful in reading the other choices because they have the wrong mix of either sunlight or sun spots, so they are incorrect.

T3C10 Which of the following bands may provide long distance communications during the peak of the sunspot cycle?

A. 6 or 10 meter bands
B. 23 centimeters
C. 70 centimeters or 1.25 meters
D. All of these choices are correct

The 23-cm, 70-cm, and 1.25-m bands all propagate over line-of-sight links, so they do not support long-distance communications. This makes Answer B, C, and D incorrect choices. During peak sunspot cycles, we can find long distance communications on the 6-m and 10-m bands, so **Answer A** is the correct choice. Don't let this question discourage you from trying those bands during sunspot minimum years as well. Propagation openings on these bands can occur at any time, especially on the 6-m band during the summer months.

T3C11 Why do VHF and UHF radio signals usually travel somewhat farther than the visual line of sight distance between two stations?

A. Radio signals move somewhat faster than the speed of light
B. Radio waves are not blocked by dust particles
C. The Earth seems less curved to radio waves than to light
D. Radio waves are blocked by dust particles

This question is asking about the radio horizon again. At radio frequencies, the Earth's atmosphere makes the Earth "look" about 4/3 bigger, so the horizon is further away, and **Answer C** is the right choice. At frequencies amateurs use, dust particle size is not important, so Answers B and D are incorrect. As we saw earlier, radio waves only travel at the speed of light, so Answer A cannot be correct.

Chapter 4

T4 – AMATEUR RADIO PRACTICES AND STATION SETUP

4.1 Introduction

In the last chapter, we saw how the Radio Frequency (RF) waves travel between stations. In this chapter, we will go deeper into the operations of an amateur radio station. The questions in this chapter will introduce you to common terminology found in amateur radio plus features of typical radio equipment the operators use. If this seems confusing, this might be time to start working with an amateur radio friend who can show you some of their equipment in context. This *Amateur Radio Practices and Station Setup* subelement has the following question groups:

A. Station setup
B. Operating controls

This will generate two questions on the Technician examination.

4.2 Radio Engineering Concepts

Shack Configuration The modern amateur radio station can be a very simple setup through a complicated demonstration of modern electronics. Figure 4.1 illustrates a representative modern shack with many of the typical components operators have for their operations. The figure centers on the shack's transmitter/receiver, also known as the *transceiver*, as the main piece of radio hardware. The shack components include

Transceiver — Radio equipment for converting the user's input/output information to and from radio signals

Antenna — The radio device to convert electrical signals to and from radio waves

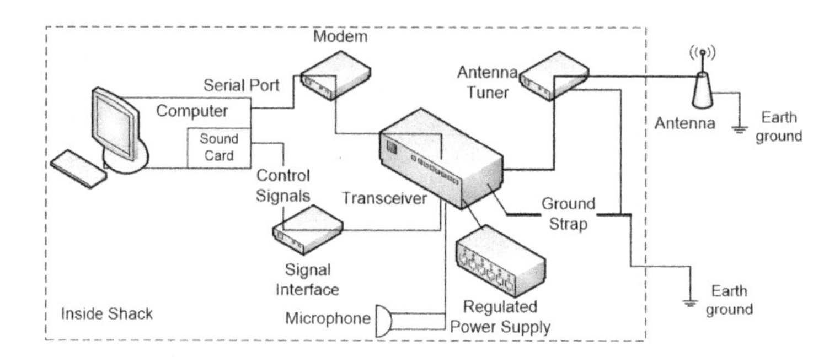

Figure 4.1: A representative modern amateur radio station configuration.

Antenna Tuner — An electronic device to match the electrical input/output of the transceiver to the antenna

Regulated Power Supply — An electronic device to give clean direct current power to the transceiver and other devices

Grounding Strap — Provides a common ground point for multiple devices as a means of ridding the system of unwanted noise and stray RF signals

Earth Ground — Tie to the Earth for unwanted, stray signals

Microphone — Permits the user to have phone communications

Computer — Permits the user to have digital communications and assist in shack operations (contact logging, Internet access, etc.)

Modem — Permits the user to interface text-type communications with the transceiver (RTTY, APRS, etc.) via the serial port interface

Signal Interface — Permits the user to interface modern digital communications needing a computer sound card (PSK, JT65, etc.) via interfaces such as USB

Figure 4.1 does not show one of the oldest components, a Morse code key, but many modern amateurs also use their sound card for CW operations.

The Transceiver A modern High Frequency (HF) transceiver does more than merely generate radio signals. Figure 4.2 illustrates the front panel for a typical transceiver. This transceiver covers 160-meter through 2-meter bands, and supports Frequency Modulation (FM), Amplitude Modulation (AM), Single Sideband (SSB), both Lower Side Band (LSB) and Upper Side Band (USB), and Continuous Wave (CW) modes. The electronics in the transceiver provide many functions to support these communication modes. Common parts of a transceiver include

Main Operating Display — Visual display of the currently tuned frequency, operating mode (FM, SSB, CW, etc.), filter bandwidth

Band Select Keys — Select operating band among allowed amateur radio bands

Variable Frequency Oscillator Control — The "tuning knob" to set the operating frequency within the selected band

Signal Strength Meter — A visual indication of the approximate received signal strength

(a) Typical transceiver and associated equipment for an amateur radio station.

(b) The main operating display for the transceiver. Top row: antenna, sideband, and filter indicators; Second row: operating frequency; Third row: received signal strength meter; Fourth row: transmit power indicator; Fifth row: Automatic Level Control indicator; Sixth row: SWR meter.

Figure 4.2: The transceiver for an amateur radio station.

Automatic Level Control — A visual display indication of how much the transmitter is adjusting the input signal level
Microphone Input — The jack for plugging in the microphone
Squelch Level Setting — Set the squelch turn-on threshold level
Receiver Incremental Tuning — Sets the receiver tuning slightly different (few Hertz) from the value indicated to assist in fine tuning
Audio Output Volume — Controls the loudness of the audio output
Memory Keys — Keys to recall pre-saved operating configurations
Control Function Keys — Soft keys for configuring the transceiver options
Naturally, this illustration is for a specific manufacturer. Other manufacturers have similar functions, but the layout will differ.

Cabling As you may have guessed from Figure 4.1, most amateur stations need a variety of cables to make the connections. In addition to the physical cables, you will also encounter a variety of cable connections that are unique to specific functions. We can group the cable and connector functions into four general categories:
RF Connectors — connectors used to carry RF signals between the rig and the antenna, and any intermediate components
Microphone Connectors — connectors to attach a microphone to the rig
Audio Connectors — connectors to attach speakers and headphones to the rig
Data Connectors — connectors to allow the rig to interface with a computer

Figure 4.3 illustrates many of the common connector types that you can find in an amateur radio shack. This is an evolving mix of connectors. With more devices supporting Wi-Fi and Bluetooth wireless connections, many of these non-RF connectors may disappear in the future.

4.3 T4A – Station Setup

4.3.1 Overview

The *Station Setup* question group in Subelement T4 concentrates on the technical aspects of setting up your amateur radio station. The *Station Setup* group covers topics such as

- Connecting microphones
- Reducing unwanted emissions
- Power source
- Connecting a computer
- RF grounding
- Connecting digital equipment
- Connecting an SWR meter

There is a total of 11 questions in this group of which one will be selected for the exam.

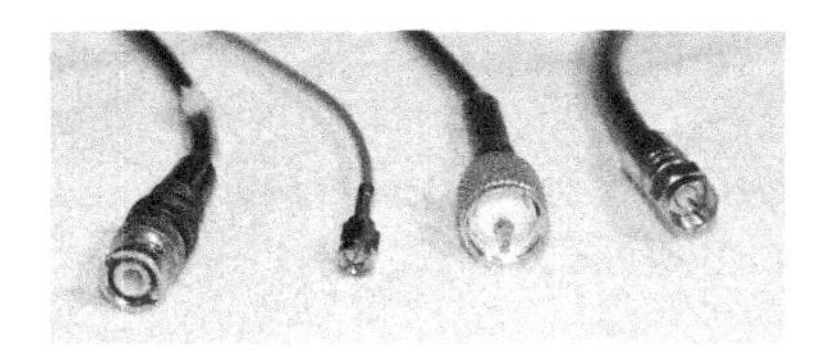

(a) BNC, SMA, UHF (PL-259), and F connectors (left to right).

(b) Various microphone connectors.

(c) RCA, ¼-inch, and 3.5 mm audio phone plugs (left to right).

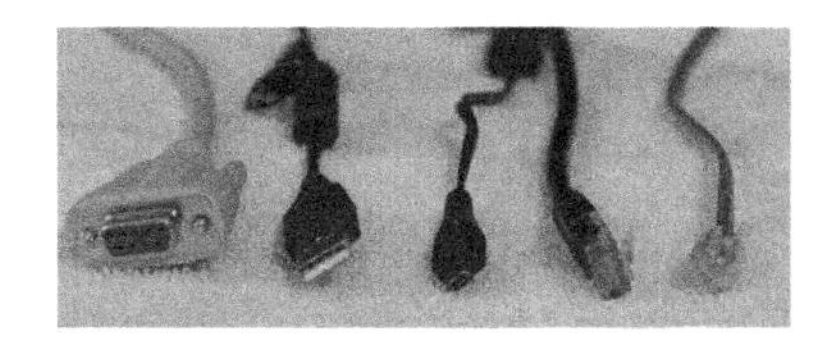

(d) 9-pin serial (DE-9), USB, mini USB, networking (RJ45), and telephone (RJ11) plugs (left to right).

Figure 4.3: Pictures of connectors used between rigs and support devices in the amateur radio shack.

4.3.2 Questions

T4A01 What must be considered to determine the minimum current capacity needed for a transceiver power supply?

A. Efficiency of the transmitter at full power output
B. Receiver and control circuit power
C. Power supply regulation and heat dissipation
D. All of these choices are correct

Designers use each of the factors listed in Answers A, B, and C when determining the current capacity for the power supply. This makes **Answer D** the best choice.

T4A02 How might a computer be used as part of an amateur radio station?

A. For logging contacts and contact information
B. For sending and/or receiving CW
C. For generating and decoding digital signals
D. All of these choices are correct

Each of the activities listed in Answers A, B, and C is a potential computer use. This makes **Answer D** the best choice for this question.

T4A03 Why should wiring between the power source and radio be heavy-gauge wire and kept as short as possible?

A. To avoid voltage falling below that needed for proper operation
B. To provide a good counterpoise for the antenna
C. To avoid RF interference
D. All of these choices are correct

If the wire is heavy gauge and short, the wire's resistance will be very small. This will keep the voltage drop across the line small as in **Answer A**. This is not for antenna considerations, so Answers B and C are not relevant. Since Answers B and C are incorrect, Answer D is also incorrect.

T4A04 Which computer sound card port is connected to a transceiver's headphone or speaker output for operating digital modes?

A. Headphone output
B. Mute
C. Microphone or line input
D. PCI or SDI

The question is asking about the audio line going from the transceiver output to the computer input. The computer sound card has two inputs: line-in or microphone as in **Answer C**. Answer A would connect the transceiver output to the sound card output, so this is incorrect. The other choices are distractions.

T4A05 What is the proper location for an external SWR meter?

A. In series with the feed line, between the transmitter and antenna
B. In series with the station's ground
C. In parallel with the push-to-talk line and the antenna
D. In series with the power supply cable, as close as possible to the radio

The Standing Wave Ratio (SWR) meter measures the power coming back from the antenna to the transmitter, so placing it in series there, as indicated in **Answer A**, is the correct engineering practice. The other choices are bad engineering distractions.

T4A06 Which of the following connections might be used between a voice transceiver and a computer for digital operation?

A. Receive and transmit mode, status, and location
B. Antenna and RF power
C. Receive audio, transmit audio, and push-to-talk (PTT)
D. NMEA GPS location and DC power

This question is asking about the interface used when you use sound cards to process the signals for modern digital modes. The three lines used are receive audio, transmit audio and Push to Talk (PTT) as in **Answer C**. The other combinations do not obtain the right mix of lines to make this work, so they are incorrect.

T4A07 How is a computer's sound card used when conducting digital communications using a computer?

A. The sound card communicates between the computer CPU and the video display
B. The sound card records the audio frequency for video display
C. The sound card provides audio to the microphone input and converts received audio to digital form
D. All of these choices are correct

The video card in your computer connects the CPU with the video display, so Answer A is incorrect. Video does not normally directly display audio, so Answer B is also incorrect. Since these two are incorrect, Answer D must also be incorrect. The audio processing in **Answer C** (see Figure 4.1) is the correct choice.

T4A08 Which of the following conductors provides the lowest impedance to RF signals?

A. Round stranded wire
B. Round copper-clad steel wire
C. Twisted-pair cable
D. Flat strap

You can make a good RF ground from a low-inductance, high current capacity conductor like a flat strip of copper. This makes **Answer D** the correct choice here. This is the ground strap in Figure 4.1. The other wires do not have the current-carrying capacity and/or the low-inductance property, so they are incorrect choices.

T4A09 Which of the following could you use to cure distorted audio caused by RF current flowing on the shield of a microphone cable?

A. Band-pass filter
B. Low-pass filter
C. Preamplifier
D. Ferrite choke

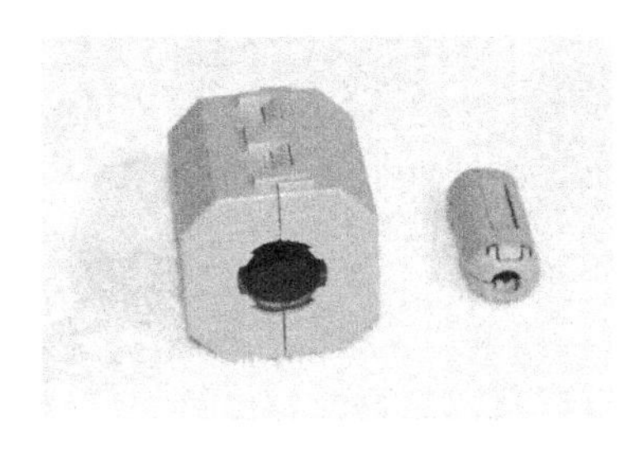

Figure 4.4: Examples of ferrite cores used to reduce noise in cables

Those small blocks you sometimes see wrapped around an electronic cable are frequently pieces of ferrite material also known as ferrite chokes, as in **Answer D**. They reduce stray RF-type signals in the electronics. The filters in Answers A and B will not reduce the RF current, so they are not good choices. A preamplifier will make a signal larger, which is probably not something you wish to do with a stray RF current.

T4A10 What is the source of a high-pitched whine that varies with engine speed in a mobile transceiver's receive audio?

A. The ignition system
B. The alternator
C. The electric fuel pump
D. Anti-lock braking system controllers

Operators frequently call this whine "alternator whine," so **Answer B** is the right choice. The other devices will not be a source of whine into a mobile transceiver.

T4A11 Where should the negative return connection of a mobile transceiver's power cable be connected?

A. At the battery or engine block ground strap
B. At the antenna mount
C. To any metal part of the vehicle
D. Through the transceiver's mounting bracket

Your mobile rig needs a good, solid ground, so you need to ground it either to the battery directly or the engine block, as in **Answer A**. The others will not guarantee a good ground for your rig.

4.4 T4B – Operating Controls

4.4.1 Overview

The *Operating Controls* question group in Subelement T4 concentrates on common aspects of the radio transmitters and receiver in your amateur radio station. The *Operating Controls* group covers topics such as

- Tuning
- Use of filters
- Squelch function
- AGC
- Transceiver operation
- Memory channels

There is a total of 13 questions in this group of which one will be selected for the exam.

4.4.2 Questions

T4B01 What may happen if a transmitter is operated with the microphone gain set too high?

A. The output power might be too high
B. The output signal might become distorted
C. The frequency might vary
D. The SWR might increase

Be careful with Answer A because it might seem like a logical result to having the microphone gain set too high. However, that is not the way radio equipment works, so this is a bad choice. **Answer B** has the actual result where the gain may cause the signal to become distorted. Answer C is also misleading because all FM transmissions have a varying carrier frequency, but that is not what the question is testing here. Answer D is not possible based on the way radios work either.

T4B02 Which of the following can be used to enter the operating frequency on a modern transceiver?

A. The keypad or VFO knob
B. The CTCSS or DTMF encoder
C. The Automatic Frequency Control
D. All of these choices are correct

The keypad and Variable Frequency Oscillator (VFO) knob are two correct ways, so **Answer A** is the best choice to answer this question. The Continuous Tone-Coded Squelch System (CTCSS) and Dual Tone Multifrequency (DTMF) electronics generate tones, but not for entering the operating frequency, so they are bad choices. The Automatic Frequency Control will help in maintaining tuning, but operators do not use it to set the frequency, so it is also a bad choice. Since Answers B and C are incorrect, Answer D must also be incorrect.

T4B03 What is the purpose of the squelch control on a transceiver?

A. To set the highest level of volume desired
B. To set the transmitter power level
C. To adjust the automatic gain control
D. To mute receiver output noise when no signal is being received

A squelch circuit mutes the output of the receiver unless the input signal is greater than a set threshold value. This keeps the receiver from trying to process noise. **Answer D** gives the correct operating function of the squelch circuit. The other choices have nothing to do with the operation of the squelch circuit.

T4B04 What is a way to enable quick access to a favorite frequency on your transceiver?

A. Enable the CTCSS tones
B. Store the frequency in a memory channel
C. Disable the CTCSS tones
D. Use the scan mode to select the desired frequency

Modern transceivers usually have a memory function to store favorite settings and frequencies. This makes **Answer B** the right choice to answer this question. Enabling and disabling the CTCSS tones will not affect access to the frequencies stored in the transceiver. Operators use them to control access to repeaters and other radios, so Answers A and C are incorrect. Answer D will help you to find the desired frequency eventually. However, this is not a good choice because the question is asking for a "quick" access method.

T4B05 Which of the following would reduce ignition interference to a receiver?

A. Change frequency slightly
B. Decrease the squelch setting
C. Turn on the noise blanker
D. Use the RIT control

The ignition interference covers a wide frequency band, so changing the frequency slightly, as in Answers A and D, will not help. Decreasing the squelch will not help because it will let in more noise, so Answer B is not a good choice. Turning on the noise blanking electronics, as in **Answer C**, is the best option of those given here.

T4B06 Which of the following controls could be used if the voice pitch of a single-sideband signal seems too high or low?

A. The AGC or limiter
B. The bandwidth selection
C. The tone squelch
D. The receiver RIT or clarifier

In SSB, if the tuning is not exactly on frequency, the voice will not be of good quality (think Donald Duck). Operators can use the receiver's Receiver Incremental Tuning (RIT) or clarifier control to make fine adjustments, so **Answer D** is the right choice. The Automatic Gain Control (AGC) or limiter in Answer A will prevent the voice from being too high in amplitude, but it will not help the pitch distortion. The bandwidth selection may make the voice totally unpleasant if the operator changes it from voice to CW, so this is not the best choice. The squelch will not affect the voice quality coming out of the speaker, so this is not a good choice.

T4B07 What does the term "RIT" mean?

A. Receiver Input Tone
B. Receiver Incremental Tuning
C. Rectifier Inverter Test
D. Remote Input Transmitter

The previous question used the term RIT, so we should be able to spot RIT is an abbreviation for Receiver Incremental Tuning, which makes **Answer B** the right choice. Be sure to read the responses carefully, so that you do not pick a close, but wrong version on the exam.

T4B08 What is the advantage of having multiple receive bandwidth choices on a multimode transceiver?

A. Permits monitoring several modes at once
B. Permits noise or interference reduction by selecting a bandwidth matching the mode
C. Increases the number of frequencies that can be stored in memory
D. Increases the amount of offset between receive and transmit frequencies

Voice, CW, and digital data sent via AM or FM all require different transmission bandwidths. The reception has the least noise when the receiver's filter bandwidth matches the transmission bandwidth. **Answer B** has the right approach: match the bandwidth to the transmission mode. The other choices are not real characteristics of adjustable receiver bandwidth, so they are incorrect choices.

T4B09 Which of the following is an appropriate receive filter bandwidth to select in order to minimize noise and interference for SSB reception?

A. 500 Hz
B. 1000 Hz
C. 2400 Hz
D. 5000 Hz

Here we need to know the approximate bandwidths of common transmission modes. The 500 Hz of Answer A is appropriate for CW, so it is not good for SSB phone. The 2400 Hz of **Answer C** is the correct choice for SSB phone. The 1000 Hz will distort the voice and the 5000 Hz is for Dual Sideband (DSB) phone.

T4B10 Which of the following is an appropriate receive filter bandwidth to select in order to minimize noise and interference for CW reception?

A. 500 Hz
B. 1000 Hz
C. 2400 Hz
D. 5000 Hz

We just saw that 2400 Hz is good for SSB phone and 5000 Hz for DSB, so Answers C and D are incorrect. The 500 Hz of **Answer A** is the correct choice.

T4B11 What is the function of automatic gain control or AGC?

A. To keep received audio relatively constant
B. To protect an antenna from lightning
C. To eliminate RF on the station cabling
D. An asymmetric goniometer control used for antenna matching

The AGC keeps the received audio as constant as possible, so **Answer A** is the right choice. Answers B and C are technically incorrect for this question. Answer D is technobabble (the goniometer is a device to measure angles and has nothing to do with antenna matching).

T4B12 Which of the following could be used to remove power line noise or ignition noise?

A. Squelch
B. Noise blanker
C. Notch filter
D. All of these choices are correct

A noise blanker removes this type of wide-band noise, so Answer B is the correct choice. Squelch sets the minimum input signal to respond to making Answer A incorrect. A notch filter removes narrow-band noise and interference and not wide-band noise, so Answer C is incorrect. Because Answers A and C are incorrect, Answer D is also incorrect.

T4B13 (C) Which of the following is a use for the scanning function of an FM transceiver?

A. To check incoming signal deviation
B. To prevent interference to nearby repeaters
C. To scan through a range of frequencies to check for activity
D. To check for messages left on a digital bulletin board

The scan function scans through the transceiver's frequency band for activity making **Answer C** correct. The other options would be nice to have, but the scan function does not do them, so they are distractions.

Chapter 5

T5 – ELECTRICAL PRINCIPLES

5.1 Introduction

In this chapter, we will begin to see questions that quiz you on fundamental concepts and terms found in radio electronics. This builds on basic concepts from electronics and the associated physics. You will need these fundamental skills to understand how radio circuits perform their jobs. As you advance in your amateur radio skill, these concepts will assist you as you build on them for your General and Extra licenses. This *Electrical Principles* subelement has the following question groups:

A. Electrical principles, units, and terms
B. Math for electronics
C. Electronic principles
D. Ohm's Law

This will generate four questions on the Technician examination.

5.2 Radio Engineering Concepts

Electrical Quantities Radio technology uses several key fundamental characteristics of circuits and the flow of energy within those circuits. We will use these quantities in this chapter and those that follow. Table 5.1 lists the properties we need to understand and master along with their units and reference symbols:
Current — the flow of electrons through a point
Voltage — the electrical potential between points
Resistance — the opposition to the flow of current
Capacitance — the ability to store energy in an electric field
Inductance — the ability to store energy in a magnetic field
Power — the rate of energy usage

Table 5.1: Fundamental Electrical Quantities

Property	Description	Unit	Symbol
Current	The flow of electrons through a point	Ampere	A
Voltage	The electrical potential between points	Volt	V
Resistance	The opposition to the flow of current	Ohm	Ω
Capacitance	The ability to store energy in an electric field	Farad	F
Inductance	The ability to store energy in a magnetic field	Henry	H
Power	The rate of energy usage	Watt	W

We will need these concepts here and in later chapters.

The electrical current flows over a *conductor* such as a wire. The conductor presents very little resistance to the electron's flow, so that it can deliver the electrical energy to its destination. In contrast to a conductor, an *insulator* prevents current flow. This is a good property for the material to wrap wires to keep them from harming users or causing short circuits.

The Electromotive Force (EMF) forces the electrons along the conductor. We measure the strength of that force between two points as a potential voltage. Devices, such as a battery that has a non-zero voltage across its terminals, produces the EMF.

Power measures the rate of energy usage in a circuit. There are many questions in this chapter using the fundamental power equation: $Power = Voltage \times Current$.

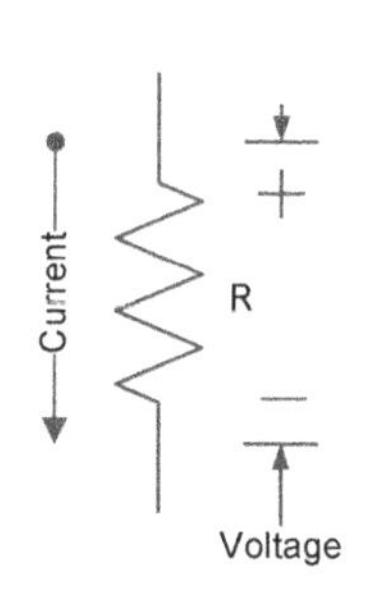

Figure 5.1: Ohm's Law relationship between V, I, and R.

Ohm's Law The fundamental electrical relationship between the voltage across a device, the current through the device, and the resistance of the device is *Ohm's Law*. Figure 5.1 shows this in a circuit environment. We will use this relationship on all Amateur Radio Service licensing exams in some form, so it is a good to learn it now. The exam questions will use all three versions of the relationship from one of the following equations:

Voltage Form — $Voltage = Resistance \times Current$

Current Form — $Current = Voltage \div Resistance$

Resistance Form — $Resistance = Voltage \div Current$

For example, a voltmeter will show that a 0.5 A current passing through a 440 Ω resistor will have a voltage of 220 V across the resistor. As we advance in the license studying, we will see how to generalize this relationship to include inductors and capacitors as well.

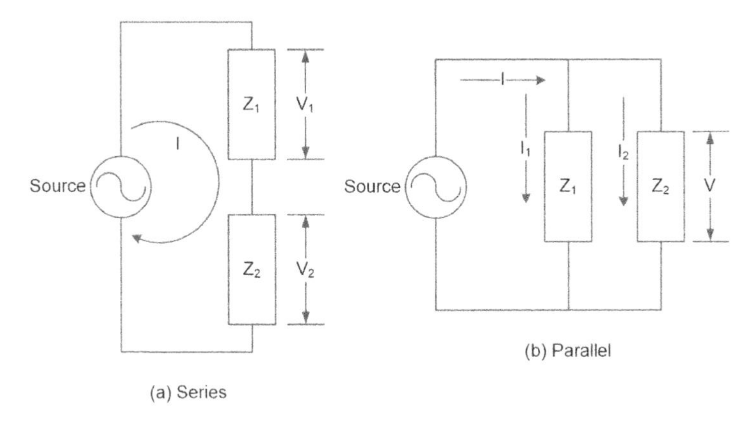

Figure 5.2: Series and parallel circuit configurations

Series and Parallel Circuits There are two important circuit configurations that you need to learn: series and parallel. We define series and parallel elements as

Series — circuit elements in series have the same current flowing through them
Parallel — circuit elements in parallel have the same voltage across them

Figure 5.2 illustrates the series and parallel configuration. Engineers sometimes call the current, I, that flows through the series elements a *mesh current*. The total current in the parallel circuit is composed of two *branch currents* that flow through the parallel circuit elements.

Unit Conversions When we apply the units found in Table 5.1, we often find the base unit is either too small or too large for our specific needs. There are standard prefixes to indicate the relative sizes of the units. Table 5.2 lists these prefixes and they are factors of 10 apart. To perform the conversion from one prefix value to another, the user either multiplies or divides by the correct factor of 10. There are many questions exercising this process.

For example, 1 A is the same as 1000 mA. We use the factor of 1000 to convert between the unit scales.

Decibels Engineers frequently use the decibel (dB) representation of a power ratio as a means of simplifying electronics computations. We compute the decibel power ratio using $Power(dB) = 10 \log (DevicePower \div ReferencePower)$. For example, we compute power relative to 1 W as $P(dB) = 10 \log [(V \times I) / 1\,\mathrm{W}]$. We indicate this as dBW. If we referenced the power to 1 mW, it would be dBm. Table 5.3 lists common dB values for increasing or decreasing by common factors such as 2 or 100 relative to the reference power.

Table 5.2: Standard Unit Prefixes

Prefix	Symbol	Multiplication Factor
exe	E	$10^{18} = 1\,000\,000\,000\,000\,000\,000$
peta	P	$10^{15} = 1\,000\,000\,000\,000\,000$
tera	T	$10^{12} = 1\,000\,000\,000\,000$
giga	G	$10^{9} = 1\,000\,000\,000$
mega	M	$10^{6} = 1\,000\,000$
kilo	k	$10^{3} = 1000$
hecto	h	$10^{2} = 100$
deca	da	$10^{1} = 10$
(unit)		1
deci	d	$10^{-1} = 0.1$
centi	c	$10^{-2} = 0.01$
milli	m	$10^{-3} = 0.001$
micro	μ	$10^{-6} = 0.000\,001$
nano	n	$10^{-9} = 0.000\,000\,001$
pico	p	$10^{-12} = 0.000\,000\,000\,001$
femto	f	$10^{-15} = 0.000\,000\,000\,000\,001$
atto	a	$10^{-18} = 0.000\,000\,000\,000\,000\,001$

Table 5.3: Common Decibel Values Relative to the Reference Power

Gain Factor	dB	Loss Factor	dB
2	3	½	-3
4	6	¼	-6
10	10	0.1	-10
100	20	0.01	-20
1000	30	0.001	-30
1000000	60	0.000001	-60

5.3 T5A – Electrical Principles, Units, and Terms

5.3.1 Overview

The *Electrical Principles, Units, and Terms* question group in Subelement T5 introduces you to electrical principles found in radio circuits. The *Electrical Principles, Units, and Terms* group covers topics such as

- Current and voltage
- Conductors and insulators
- Alternating and direct current
- Series and parallel circuits

There is a total of 14 questions in this group of which one will be selected for the exam.

5.3.2 Questions

T5A01 Electrical current is measured in which of the following units?
A. Volts
B. Watts
C. Ohms
D. Amperes

This is the first of several questions using the information contained in Table 5.1. Voltage is for measuring electrical potential, watts are for measuring electrical power, ohms are for measuring electrical resistance. We measure electrical current in amperes, so **Answer D** is the correct choice.

T5A02 Electrical power is measured in which of the following units?
A. Volts
B. Watts
C. Ohms
D. Amperes

Using the answer to the last question as a guide, you should be able to spot watts in **Answer B** as the correct choice here.

T5A03 What is the name for the flow of electrons in an electric circuit?
A. Voltage
B. Resistance
C. Capacitance
D. Current

Electrical current is the flow of electrons in a device, so **Answer D** is the right choice. Voltage is the strength of an electrical potential between two points. Resistance is a restriction to the flow of electrical current, so it is related, but not the exact definition, making Answer B an incorrect choice. Capacitance is the ability to store energy in an electric field, which makes Answer C incorrect as well.

T5A04 What is the name for a current that flows only in one direction?
A. Alternating current
B. Direct current
C. Normal current
D. Smooth current

Direct Current (DC) flows in one direction, so **Answer B** is the correct choice. Alternating Current (AC) flows in two directions, so Answer A is incorrect. Answers C and D are silly distractions

T5A05 What is the electrical term for the electromotive force (EMF) that causes electron flow?
A. Voltage
B. Ampere-hours
C. Capacitance
D. Inductance

The EMF is another name for electrical potential, and either name is the reason behind the flow of electrons, so the choice of voltage in **Answer A** is correct. Ampere-hours tell you how much charge the battery has stored. Capacitance is the ability to sore energy in an electric field. Inductance is the ability to store energy in a magnetic field. These are not what the question is asking for.

T5A06 How much voltage does a mobile transceiver usually require?
A. About 12 volts
B. About 30 volts
C. About 120 volts
D. About 240 volts

If you think about this question, the answer should be obvious: your mobile rig lives in your car, and your car typically has a 12-volt battery, so the rig ought to require 12 Volts, as in **Answer A**. Answers C and D have the voltages for AC power that you might find in your house, so they are incorrect here.

T5A07 Which of the following is a good electrical conductor?
A. Glass
B. Wood
C. Copper
D. Rubber

A good conductor allows electrical current to flow easily and without much loss. Of these choices, copper, as in **Answer C**, is the material having the best conduction properties. The other choices are all insulators that block the flow of electrical current.

T5A08 Which of the following is a good electrical insulator?
A. Copper
B. Glass
C. Aluminum
D. Mercury

This is the opposite of the previous question. Here, glass is the only insulator of the choices given, so **Answer B** is the right choice to answer this question. The other choices are all conductors, so they do not make good electrical insulators.

T5A09 What is the name for a current that reverses direction on a regular basis?
A. Alternating current
B. Direct current
C. Circular current
D. Vertical current

Earlier we had DC that flows in one direction, so Answer B is incorrect. The AC of **Answer A** is the right choice here. Answers C and D are silly distractions.

T5A10 Which term describes the rate at which electrical energy is used?
A. Resistance
B. Current
C. Power
D. Voltage

Power is a measurement of the rate at which we use electrical energy, so **Answer C** is the right choice. Resistance, current, and voltage all affect the power, but we cannot use them individually to calculate the rate of electrical energy usage, so they are not correct choices.

T5A11 What is the unit of electromotive force?
A. The volt
B. The watt
C. The ampere
D. The ohm

This question should look familiar by now, and you should be able to spot **Answer A** as the right choice because it has the unit of volts. Remember that the watt is for power, the ampere is for current, and the ohm is for resistance, so these are incorrect choices.

T5A12 What describes the number of times per second that an alternating current makes a complete cycle?

A. Pulse rate
B. Speed
C. Wavelength
D. Frequency

The number of reversals per second is the frequency, so **Answer D** is the right choice. We discuss the wavelength with the properties of Radio Frequency (RF) waves, but not normally with currents. Answers A and B are to distract you.

T5A13 In which type of circuit is current the same through all components?

A. Series
B. Parallel
C. Resonant
D. Branch

In a series circuit, the same current flows through all components as in **Answer A**. A parallel circuit has the same voltage across all circuit elements, so Answer B is for the next question. A resonant circuit acts like a resistive circuit at the AC frequency, so this is incorrect here. A branch current is a current through a selected set of circuit elements, so it is not as good a choice as Answer A.

T5A14 In which type of circuit is voltage the same across all components?

A. Series
B. Parallel
C. Resonant
D. Branch

As we just saw in the previous question, series current flow through all circuit elements and parallel circuits have the same voltage across all elements. This makes Answer B the correct choice here. Resonant and branch currents are still distractions here.

5.4 T5B – Math for Electronics

5.4.1 Overview

The *Math for Electronics* question group in Subelement T5 introduces you to quantities found in radio circuits in various units of measurement. The *Math for Electronics* group covers topics such as

- Conversion of electrical units
- Decibels
- The metric system

There is a total of 13 questions in this group of which one will be selected for the exam.

5.4.2 Questions

T5B01 How many milliamperes is 1.5 amperes?

A. 15 milliamperes
B. 150 milliamperes
C. 1,500 milliamperes
D. 15,000 milliamperes

Here, we start a series of questions where we need to convert units. The factors listed in Table 5.2 will be helpful here. To convert from amperes to milliamperes, multiply the quantity by 1000. 1500 mA is the same as 1.5 A, as in **Answer C**. The other choices are all math mistakes: Answer A multiplies by 10, Answer B multiplies by 100, and Answer D multiplies by 10 000.

T5B02 What is another way to specify a radio signal frequency of 1,500,000 hertz?

A. 1500 kHz
B. 1500 MHz
C. 15 GHz
D. 150 kHz

To express the frequency in kilohertz (kHz), we divide by 1000, so 1 500 000 Hz = 1500 kHz. To express the answer in megahertz (MHz), we divide by 1 000 000, so 1 500 000 Hz = 1.5 MHz. To express the frequency in gigahertz (GHz), we divide by 1 000 000 000, so 1 500 000 Hz = 0.0015 GHz. As we can see, **Answer A** is the correct choice.

T5B03 How many volts are equal to one kilovolt?

A. One one-thousandth of a volt
B. One hundred volts
C. One thousand volts
D. One million volts

Here we use the same principles as before, but now with voltage: 1000 V = 1 kV. This makes **Answer C** the correct choice. The other options multiply or divide by the wrong factors of 10.

T5B04 How many volts are equal to one microvolt?

A. One one-millionth of a volt
B. One million volts
C. One thousand kilovolts
D. One one-thousandth of a volt

This is a new prefix: one micro implies one millionth of the base unit. Therefore, there are 1 000 000 µV in 1 V or 0.000 001 V in 1 µV. **Answer A** is the right choice. Note: be careful in your general reading because sometimes the fonts get changed, and you will see µV printed as "uV."

T5B05 Which of the following is equivalent to 500 milliwatts?

A. 0.02 watts
B. 0.5 watts
C. 5 watts
D. 50 watts

To convert from milliwatts to watts, we divide the quantity by 1000, so 500 mW = 0.5 W. This makes **Answer B** the correct choice.

T5B06 If an ammeter calibrated in amperes is used to measure a 3000-milliampere current, what reading would it show?

A. 0.003 amperes
B. 0.3 amperes
C. 3 amperes
D. 3,000,000 amperes

You should be able to work this one by now: 3000 mA = 3 A (divide the measure in milliamps by 1000 to get amperes). You should have picked **Answer C** as the correct choice.

T5B07 If a frequency readout calibrated in megahertz shows a reading of 3.525 MHz, what would it show if it were calibrated in kilohertz?

A. 0.003525 kHz
B. 35.25 kHz
C. 3525 kHz
D. 3,525,000 kHz

To get kilohertz from megahertz, multiply by 1000: 3.525 MHz = 3525 kHz, which makes **Answer C** the correct choice.

T5B08 How many microfarads are 1,000,000 picofarads?

A. 0.001 microfarads
B. 1 microfarad
C. 1000 microfarads
D. 1,000,000,000 microfarads

Another new unit: pico is one trillionth of the base unit. Between micro and pico, there is a scale factor of 1 000 000. This makes 1 000 000 pF = 1 µF, and **Answer B** is the right choice.

T5B09 What is the approximate amount of change, measured in decibels (dB), of a power increase from 5 watts to 10 watts?

A. 2 dB
B. 3 dB
C. 5 dB
D. 10 dB

For these next questions, Table 5.3 may be useful. The decibel unit is a logarithmic measure of power relative to a reference power. In this case, the increase is $10\log(10\,\mathrm{W}/5\,\mathrm{W}) = 10\log(2) = 3\,\mathrm{dB}$. This makes **Answer B** the correct choice. Answer A represents a change from 5 W to 7.9 W. Answer C represents a change from 5 W to 15.8 W. Answer D represents a change from 5 W to 50 W.

T5B10 What is the approximate amount of change, measured in decibels (dB), of a power decrease from 12 watts to 3 watts?

A. -1 dB
B. -3 dB
C. -6 dB
D. -9 dB

Using the previous question as a guide, the decrease would be $10\log(3\,\mathrm{W}/12\,\mathrm{W}) = 10\log(0.25) = -6\,\mathrm{dB}$. A negative value means that there was a decrease. From this, we see that **Answer C** is correct. Answer A corresponds to a power decrease to 9.53 W, Answer B corresponds to a power decrease to 6 W, while Answer D corresponds to a decrease to 1.5 W.

T5B11 What is the approximate amount of change, measured in decibels (dB), of a power increase from 20 watts to 200 watts?

A. 10 dB
B. 12 dB
C. 18 dB
D. 28 dB

This is an easy one! Factor of 10 changes are always 10 dB changes (try out the equation). Changing from 20 W to 200 W is a factor of 10 increase, so that is a 10-dB increase, as in **Answer A**.

T5B12 Which of the following frequencies is equal to 28,400 kHz?

A. 28.400 MHz
B. 2.800 MHz
C. 284.00 MHz
D. 28.400 kHz

To get megahertz from kilohertz, divide by 1000: 28 400 kHz = 28.400 MHz, which makes **Answer A** the correct choice.

T5B13 If a frequency readout shows a reading of 2425 MHz, what frequency is that in GHz?

A. 0.002425 GHZ
B. 24.25 GHz
C. 2.425 GHz
D. 2425 GHz

To get gigahertz from megahertz, divide by 1000: 2425 MHz = 2.425 GHz, which makes **Answer C** the correct choice.

5.5 T5C - Electronic Principles

5.5.1 Overview

The *Electronic Principles* question group in Subelement T5 continues with more components in radio circuits, and the various units of measurement. The *Electronic Principles* group covers topics such as

- Capacitance
- Inductance
- Current flow in circuits
- Alternating current
- Definition of RF
- Definition of polarity
- DC power calculations
- Impedance

There is a total of 14 questions in this group of which one will be selected for the exam.

5.5.2 Questions

T5C01 What is the ability to store energy in an electric field called?

A. Inductance
B. Resistance
C. Tolerance
D. Capacitance

The ability to store energy in an electric field is capacitance, as we saw in Table 5.1, which makes **Answer D** the correct choice. Inductance is storing energy in a magnetic field, and resistance is the opposition to the flow of current. Manufacturers use tolerance to indicate how much variation in a component's value they expect to see, and it is not a measurement of electrical energy storage.

T5C02 What is the basic unit of capacitance?

A. The farad
B. The ohm
C. The volt
D. The henry

We use farads to measure capacitance, which makes **Answer A** is the correct choice. Resistance is in ohms, electrical potential is in volts, and inductance is in henrys.

T5C03 What is the ability to store energy in a magnetic field called?

A. Admittance
B. Capacitance
C. Resistance
D. Inductance

You should have spotted **Answer D**, inductance, as the right choice here. The new term, admittance, is the ease of current flow, and is the reciprocal of resistance.

T5C04 What is the basic unit of inductance?

A. The coulomb
B. The farad
C. The henry
D. The ohm

The basic unit of inductance is the henry, as in **Answer C**. The coulomb is the measure of charge, farad is the measure of capacitance, and ohm is the measure of resistance.

T5C05 What is the unit of frequency?

A. Hertz
B. Henry
C. Farad
D. Tesla

The unit of frequency, as we saw earlier, is the hertz, which makes **Answer A** the correct choice. By now, you know that the henry and the farad are incorrect choices for this question. The tesla is a measure of magnetic field strength.

T5C06 What does the abbreviation "RF" refer to?

A. Radio frequency signals of all types
B. The resonant frequency of a tuned circuit
C. The real frequency transmitted as opposed to the apparent frequency
D. Reflective force in antenna transmission lines

As we have seen, Radio Frequency (RF) is for all radio frequencies of all types, so **Answer A** is the right choice. The other choices are to distract you.

T5C07 A radio wave is made up of what type of energy?
A. Pressure
B. Electromagnetic
C. Gravity
D. Thermal

A radio wave is a form of electromagnetic radiation making **Answer B** the correct choice. Sound waves are pressure waves, gravity waves are due to physical mass, and scientists have only recently proven they exist, so these are incorrect. Thermal waves are "heat waves" and a distraction for you.

T5C08 What is the formula used to calculate electrical power in a DC circuit?
A. Power (P) equals voltage (E) multiplied by current (I)
B. Power (P) equals voltage (E) divided by current (I)
C. Power (P) equals voltage (E) minus current (I)
D. Power (P) equals voltage (E) plus current (I)

This is the equation to know: $Power(P) = Voltage(V) \times Current(I)$. This makes **Answer A** the correct choice. Read the choices carefully on the test because the others have the right quantities, but with the wrong operations. Note: I find it easier to remember V, so I will use that in the equations. The exam uses E.

T5C09 How much power is being used in a circuit when the applied voltage is 13.8 volts DC and the current is 10 amperes?
A. 138 watts
B. 0.7 watts
C. 23.8 watts
D. 3.8 watts

To make this computation we don't even need a calculator to apply the formula: $P = VI$. Here, $P = 13.8\,\text{V} \times 10\,\text{A} = 138\,\text{W}$, as in **Answer A**.

T5C10 How much power is being used in a circuit when the applied voltage is 12 volts DC and the current is 2.5 amperes?
A. 4.8 watts
B. 30 watts
C. 14.5 watts
D. 0.208 watts

We use the same relationship with this question: $P = VI$. Here, $P = 12\,\text{V} \times 2.5\,\text{A} = 30\,\text{W}$, as in **Answer B**.

T5C11 How many amperes are flowing in a circuit when the applied voltage is 12 volts DC and the load is 120 watts?

A. 0.1 amperes
B. 10 amperes
C. 12 amperes
D. 132 amperes

Now we get to invert the same formula. Remember $P = VI$, so $I = P/V$. Applying the numbers, $I = 120\,\mathrm{W}/12\,\mathrm{V} = 10\,\mathrm{A}$, as in **Answer B**.

T5C12 What is impedance?

A. It is a measure of the opposition to AC current flow in a circuit
B. It is the inverse of resistance
C. It is a measure of the Q or Quality Factor of a component
D. It is a measure of the power handling capability of a component

Impedance is a term that generalizes resistance, especially for AC circuits, which makes **Answer A** the correct choice. Answer B is the admittance. Answer C is a measure of how narrow-band the circuit is, and we use it with filters and tuned circuits. Answer D is technically incorrect as well.

T5C13 What is a unit of impedance?

A. Volts
B. Amperes
C. Coulombs
D. Ohms

Although we are considering AC circuits, the impedance still has the units of ohms like resistance, so **Answer D** is the right choice. You should be able to say what measurements the other choices correspond to.

T5C14 (D) What is the proper abbreviation for megahertz?

A. mHz
B. mhZ
C. Mhz
D. MHz

The abbreviation has two parts: the scaling part and the physical units part. Upper-case and lower-case letters are important in both. Answer A corresponds to milli-Hertz, so it is incorrect. Answer B is incorrect because it has the "m" for milli and "hz" is an improper unit designation. Answer C is incorrect because it has the "mega" right, but the units are wrong. **Answer D** is correct because it has both parts right.

5.6 T5D – Ohm's Law

5.6.1 Overview

The *Ohm's Law* question group in Subelement T5 covers a fundamental electrical voltage and current relationship between electronic components used in radio circuits. The *Ohm's Law* group covers the formulas and usage of this relationship for series and parallel circuits. There is a total of 16 questions in this group of which one will be selected for the exam.

5.6.2 Questions

T5D01 What formula is used to calculate current in a circuit?
A. Current (I) equals voltage (E) multiplied by resistance (R)
B. Current (I) equals voltage (E) divided by resistance (R)
C. Current (I) equals voltage (E) added to resistance (R)
D. Current (I) equals voltage (E) minus resistance (R)

Here is the first of several questions using the Ohm's Law relationship between voltage, current, and resistance. To compute current using Ohm's Law, we use the formula $Current(I) = Voltage(E) \div Resistance(R)$, as in **Answer B**. The other choices use the same quantities, but with the wrong operations, so read the choices carefully.

T5D02 What formula is used to calculate voltage in a circuit?
A. Voltage (E) equals current (I) multiplied by resistance (R)
B. Voltage (E) equals current (I) divided by resistance (R)
C. Voltage (E) equals current (I) added to resistance (R)
D. Voltage (E) equals current (I) minus resistance (R)

Here, they are asking you about the voltage form of Ohm's Law or $Voltage(E) = Current(I) \times Resistance(R)$, as in **Answer A**. Again, be careful with the other choices on the exam because they all have the right quantities, but the wrong operations for computing voltage.

T5D03 What formula is used to calculate resistance in a circuit?
A. Resistance (R) equals voltage (E) multiplied by current (I)
B. Resistance (R) equals voltage (E) divided by current (I)
C. Resistance (R) equals voltage (E) added to current (I)
D. Resistance (R) equals voltage (E) minus current (I)

Finally, we look at the resistance form of Ohm's Law where $Resistance(R) = Voltage(E) \div Current(I)$. This makes **Answer B** the correct choice.

T5D04 What is the resistance of a circuit in which a current of 3 amperes flows through a resistor connected to 90 volts?

A. 3 ohms
B. 30 ohms
C. 93 ohms
D. 270 ohms

Now we get to use the resistance form of Ohm's Law: $R = V/I$. Using the quantities given in the question, $R = 90\,\text{V}/3\,\text{A} = 30\,\Omega$, as in **Answer B**.

T5D05 What is the resistance in a circuit for which the applied voltage is 12 volts and the current flow is 1.5 amperes?

A. 18 ohms
B. 0.125 ohms
C. 8 ohms
D. 13.5 ohms

You should be ready to use the same equation as for the last question: $R = V/I$. Applying the values gives $R = 12\,\text{V}/1.5\,\text{A} = 8\,\Omega$, as in **Answer C**.

T5D06 What is the resistance of a circuit that draws 4 amperes from a 12-volt source?

A. 3 ohms
B. 16 ohms
C. 48 ohms
D. 8 Ohms

Are you ready to use the equation again? $R = V/I$, so the resistance here is $12\,\text{V}/4\,\text{A}$ or $3\,\Omega$, as in **Answer A**.

T5D07 What is the current flow in a circuit with an applied voltage of 120 volts and a resistance of 80 ohms?

A. 9600 amperes
B. 200 amperes
C. 0.667 amperes
D. 1.5 amperes

Now we get to use the current version of Ohm's Law: $I = V/R$. Using the values from the question: $I = 120\,\text{V}/80\,\Omega$ or $I = 1.5\,\text{A}$. **Answer D** is the correct choice.

T5D08 What is the current flowing through a 100-ohm resistor connected across 200 volts?
A. 20,000 amperes
B. 0.5 amperes
C. 2 amperes
D. 100 amperes

We get to use the same current equation again this time, so $I = V/R$. Applying the given values: $I = 200\,\text{V}/100\,\Omega$. The 2 A in **Answer C** is the correct choice.

T5D09 What is the current flowing through a 24-ohm resistor connected across 240 volts?
A. 24,000 amperes
B. 0.1 amperes
C. 10 amperes
D. 216 amperes

Another current flow computation: $I = V/R$. Using the values gives $I = 240\,\text{V}/24\,\Omega$. Did you pick **Answer C** as correct with 10 A?

T5D10 What is the voltage across a 2-ohm resistor if a current of 0.5 amperes flows through it?
A. 1 volt
B. 0.25 volts
C. 2.5 volts
D. 1.5 volts

Now we are on the voltage form of Ohm's Law: $V = IR$. Using the value here, $V = 2\,\Omega \times 0.5\,\text{A}$. Did you need a calculator to compute 1 V, as in **Answer A**?

T5D11 What is the voltage across a 10-ohm resistor if a current of 1 ampere flows through it?
A. 1 volt
B. 10 volts
C. 11 volts
D. 9 volts

Using the voltage equation again, $V = IR$, so $V = 1\,\text{A} \times 10\,\Omega$. **Answer B** is correct with 10 V.

T5D12 What is the voltage across a 10-ohm resistor if a current of 2 amperes flows through it?

A. 8 volts
B. 0.2 volts
C. 12 volts
D. 20 volts

Here we have the last time using the voltage form of Ohm's Law: $V = IR$. Using the values given in the question, $V = 2\,\mathrm{A} \times 10\,\Omega$. This makes **Answer D**, with 20 V, the correct choice.

T5D13 What happens to current at the junction of two components in series?

A. It divides equally between them
B. It is unchanged
C. It divides based on the on the value of the components
D. The current in the second component is zero

For circuit elements in series, the same current flows through the elements. This means that the current at the junction is unchanged as in **Answer B**. Answer C is for parallel circuit elements, so it is incorrect here. Answers A and D are distractions.

T5D14 What happens to current at the junction of two components in parallel?

A. It divides between them dependent on the value of the components
B. It is the same in both components
C. Its value doubles
D. Its value is halved

In parallel circuit elements, the voltage across them is the same, but the current splits according to the value of the components. This makes **Answer A** correct. Answer B is for series circuit elements, so it is incorrect here. Answers C and D are distractions.

T5D15 What is the voltage across each of two components in series with a voltage source?

A. The same voltage as the source
B. Half the source voltage
C. It is determined by the type and value of the components
D. Twice the source voltage

While the current is the same for series elements, the voltage across each element depends on the value of the component. **Answer C** is correct in this case. The other choices are to distract you.

T5D16 (D)What is the voltage across each of two components in parallel with a voltage source?

A. It is determined by the type and value of the components
B. Half the source voltage
C. Twice the source voltage
D. The same voltage as the source

In this parallel circuit configuration, the voltage is the same across both components and the same as the source. This makes **Answer D** the correct choice. The other choices are to distract you.

Chapter 6

T6 – ELECTRICAL COMPONENTS

6.1 Introduction

The Electrical Components chapter presents questions related to discrete electrical components and introduces you to circuit diagrams containing those components. This builds on the electrical principles from Chapter 5 and applies them to real devices. This *Electrical Components* subelement has the following question groups:

A. Electrical components
B. Semiconductors
C. Circuit diagrams
D. Component functions

This will generate four questions on the Technician examination.

6.2 Radio Engineering Concepts

Electrical Components We start with the electrical components we saw earlier, namely

Resistor — A device for limiting the flow of current by using a partially conducting medium for current flow

Capacitor — A device, composed of two parallel plates separated by an insulator, for storing energy in an electric field

Inductor — A device, made by one or more coils of wire around a common center, for storing energy in a magnetic field

Each of these devices is available as a fixed-value device or a variable-value device. The variable resistor is also known as a *potentiometer*. Designers use pairs of inductors to make a *transformer*, which is a device frequently used for changing voltage levels in circuits. Designers also use circuits containing both inductors and capacitors in the right configuration to make oscillators, which are devices to produce a carrier signal, and filters to modify a signal.

New devices for this chapter are

Switch — A device for controlling the path the current flow takes by making or breaking connections

Relay — A switch having an electromagnet to control the switching action

Fuse — A device for protecting circuits by preventing a current overload

Battery — A device for storing energy to be used in running a circuit

Switches come in several varieties depending upon how many poles (the number of input lines on the switch) and how many throws (the number of output combinations each input can connect to) the switch has. Manufacturers rate fuses based on the threshold current value at which they trip and stop conducting, for example 10 A. Batteries come in two classes: *rechargable* and *non-rechargable*.

Semiconductor Components The semiconductor industry has greatly expanded the number of devices, and the number of functions that a circuit designer has available for making useful circuits in all electronic applications. Radio technology has benefited from this explosion in available components. The semiconductor devices that you will encounter here are

Diode — A device that permits current flow in a single direction

Transistor — A device that uses a voltage or a current to control current flow or amplify a signal

Regulator — A device to control the voltage coming from a power supply

Diodes have two terminals: an *anode* where positive current flows in, and a *cathode* where positive charge flows out. Diodes comes in several different formats such as light emitting or photosensitive. Diodes have two semiconductor layers that form one junction. Transistors come in two major families: *Bipolar Junction Transistor* and *Field Effect Transistor*. The bipolar transistors have two divisions: NPN and PNP based on the types of materials used. They also have two semiconductor junctions. The field effect transistors have multiple types based upon the semiconductors used in their design. Both transistor types have three terminals. The bipolar transistor terminals are emitter, base, and collector. The field effect transistor terminals are source, gate, and drain. Frequently, manufacturers package multiple semiconductor devices together in an *integrated circuit*, which produces additional capabilities than the individual devices can do on their own.

Circuit Diagrams Engineers use *circuit diagram* to illustrate how they have connected the individual components to make a functioning circuit. To assist this process, engineers use standard symbols to represent the individual devices, as Figure 6.1 illustrates. This is not a full list of symbols, but they will get you through the licensing process.

Circuit diagrams use a straight line to indicate the connections between the components. With some devices, such as semiconductors and polarized capacitors, the device will specify specific input or output terminals for making a connection. If

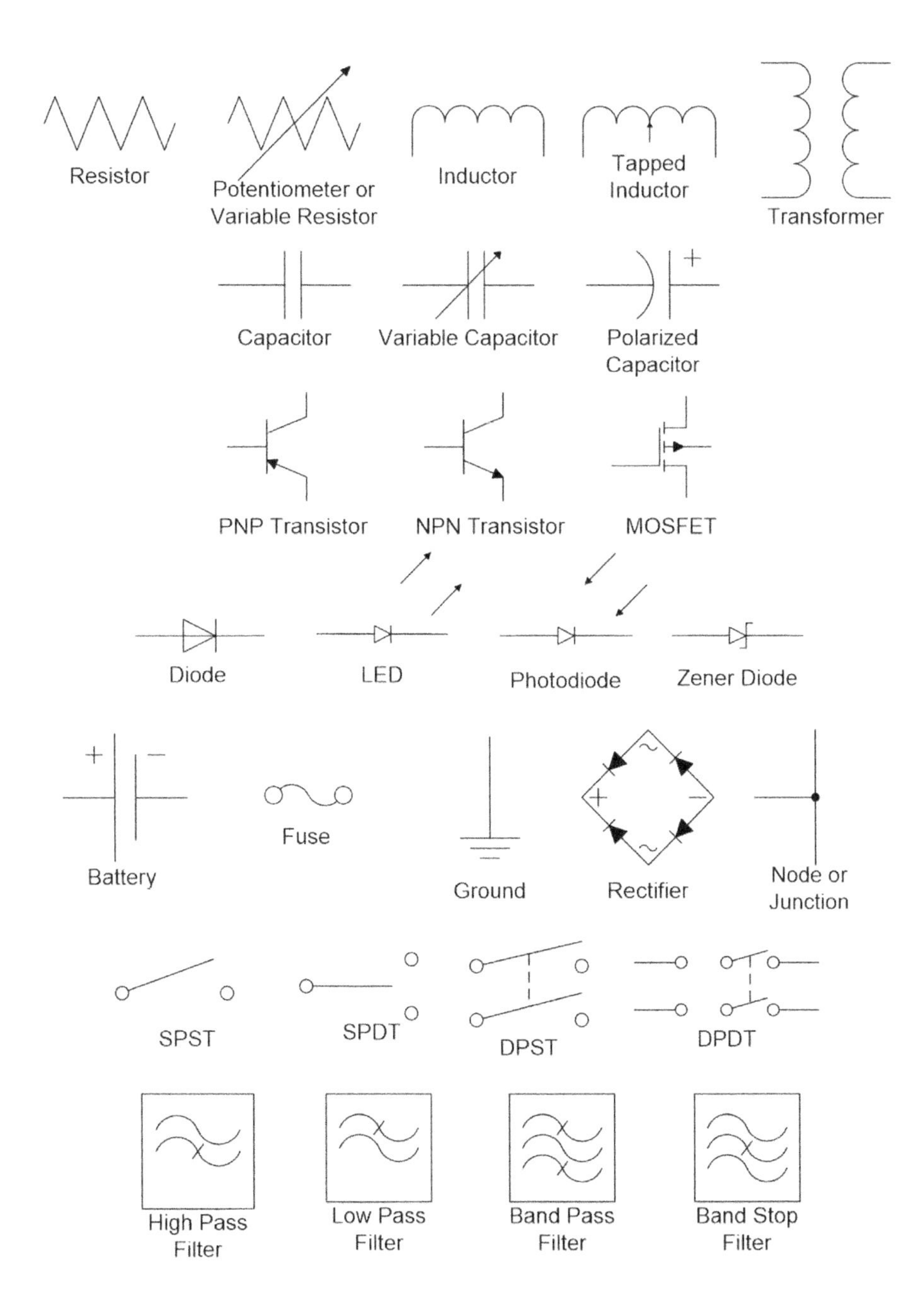

Figure 6.1: Common symbols used for circuit elements.

the user violates these rules, the device will heat up, which causes the "magic white smoke" to escape and damage the device. Also, with complicated diagrams, the connecting wires may cross each other. The general rule is that if the wires cross, but do not have a "node" symbol at the junction, then designers do not consider them to be connected. Sometimes, designers use a jumper arc to indicate no connection.

6.3 T6A – Electrical Components

6.3.1 Overview

The *Electrical Components* question group in Subelement T6 has further questions on the properties of electrical components. The *Electrical Components* group covers topics such as

- Fixed and variable resistors
- Capacitors and inductors
- Fuses
- Switches
- Batteries

There is a total of 11 questions in this group of which one will be selected for the exam.

6.3.2 Questions

T6A01 What electrical component opposes the flow of current in a DC circuit?
A. Inductor
B. Resistor
C. Voltmeter
D. Transformer

Resistors oppose the flow of current, so **Answer B** is the correct choice. A voltmeter measures electrical potential, while designers use inductors and transformers because of their magnetic field properties. Figure 6.1 includes the circuit schematic symbols for resistors.

T6A02 What type of component is often used as an adjustable volume control?
A. Fixed resistor
B. Power resistor
C. Potentiometer
D. Transformer

The volume knob on your radio is an example of a variable resistor, also known as a potentiometer, which makes **Answer C** the correct choice. A fixed resistor does not permit any audio volume adjustment, while designers do not use power resistors or transformers to make volume controls.

T6A03 What electrical parameter is controlled by a potentiometer?

A. Inductance
B. Resistance
C. Capacitance
D. Field strength

By now you should be able to spot "resistance" in **Answer B** as the correct choice because a potentiometer is a variable resistor.

T6A04 What electrical component stores energy in an electric field?

A. Resistor
B. Capacitor
C. Inductor
D. Diode

As we saw in Table 5.1, capacitors store energy in an electric field, and this makes **Answer B** the correct choice. Resistors and diodes are not energy storage devices. You need to remember that inductors store energy in a magnetic field. Figure 6.1 includes the circuit schematic symbols for capacitors.

T6A05 What type of electrical component consists of two or more conductive surfaces separated by an insulator?

A. Resistor
B. Potentiometer
C. Oscillator
D. Capacitor

A capacitor is composed of two conductive surfaces separated by an insulator, so **Answer D** is the right choice.

T6A06 What type of electrical component stores energy in a magnetic field?

A. Resistor
B. Capacitor
C. Inductor
D. Diode

As we saw in Table 5.1, you should be able to spot the inductor of **Answer C** as the right choice. The capacitor stores energy in an electric field. Resistors and diodes are not energy storage devices. Figure 6.1 includes the circuit schematic symbols for inductors.

T6A07 What electrical component usually is constructed as a coil of wire?
A. Switch
B. Capacitor
C. Diode
D. Inductor

If you look at the symbol for the inductor in Figure 6.1, you can see that it is based on a coil of wire. This makes **Answer D** the correct choice.

T6A08 What electrical component is used to connect or disconnect electrical circuits?
A. Magnetron
B. Switch
C. Thermistor
D. All of these choices are correct

The answer to this is obvious: designers use a switch for connecting and disconnecting electrical devices, which makes **Answer B** the right choice here. Figure 6.1 includes the circuit schematic symbols for switches.

T6A09 What electrical component is used to protect other circuit components from current overloads?
A. Fuse
B. Capacitor
C. Inductor
D. All of these choices are correct

The fuse, as in **Answer A**, protects circuits from an overload. An overload will usually destroy the capacitor and inductor, which is a bad thing. With Answers B and C incorrect, Answer D is incorrect.

T6A10 Which of the following battery types is rechargeable?
A. Nickel-metal hydride
B. Lithium-ion
C. Lead-acid gel-cell
D. All of these choices are correct

Each of the battery types listed in Answers A, B, and C is rechargeable, so **Answer D** is the right choice for this question.

T6A11 Which of the following battery types is not rechargeable?
A. Nickel-cadmium
B. Carbon-zinc
C. Lead-acid
D. Lithium-ion

Of the choices given, only the carbon-zinc battery of **Answer B** is not rechargeable, so it is the right choice. The others are all rechargeable battery technologies.

6.4 T6B – Semiconductors

6.4.1 Overview

The *Semiconductors* question group in Subelement T6 introduces you to semiconductor devices. The *Semiconductors* group covers topics such as

- Basic principles and applications of solid state devices
- Diodes and transistors

There is a total of 11 questions in this group of which one will be selected for the exam.

6.4.2 Questions

T6B01 What class of electronic components uses a voltage or current signal to control current flow?

A. Capacitors
B. Inductors
C. Resistors
D. Transistors

A transistor uses a voltage or a current to control the flow of current making **Answer D** the correct choice. None of the other choices have this property. Figure 6.1 includes the circuit schematic symbols for transistors.

T6B02 What electronic component allows current to flow in only one direction?

A. Resistor
B. Fuse
C. Diode
D. Driven Element

A diode permits current flow in a single direction making **Answer C** the correct choice. The resistor and the fuse permit flow in both directions. The "driven element" choice is not specific enough to know what it is doing. Figure 6.1 includes the circuit schematic symbols for diodes.

T6B03 Which of these components can be used as an electronic switch or amplifier?
A. Oscillator
B. Potentiometer
C. Transistor
D. Voltmeter

Two uses for transistors are as switches and amplifiers making **Answer C** the correct choice here.

T6B04 Which of the following components can consist of three layers of semiconductor material?
A. Alternator
B. Transistor
C. Triode
D. Pentagrid converter

A Bipolar Junction Transistor has three layers of semiconductor, so **Answer B** is the correct choice.

T6B05 Which of the following electronic components can amplify signals?
A. Transistor
B. Variable resistor
C. Electrolytic capacitor
D. Multi-cell battery

As we saw above, a transistor can amplify signal, so **Answer A** is the correct choice among those given here. The other devices do not amplify signals.

T6B06 How is the cathode lead of a semiconductor diode often marked on the package?
A. With the word cathode
B. With a stripe
C. With the letter C
D. All of these choices are correct

If you look at a diode, you will see a stripe at the one end. This marks the cathode, and **Answer B** is the right choice.

T6B07 What does the abbreviation LED stand for?
A. Low Emission Diode
B. Light Emitting Diode
C. Liquid Emission Detector
D. Long Echo Delay

A Light Emitting Diode (LED), as found in **Answer B**, is the right choice. The others are distractions to trip you up when taking the exam.

T6B08 What does the abbreviation FET stand for?
A. Field Effect Transistor
B. Fast Electron Transistor
C. Free Electron Transition
D. Frequency Emission Transmitter

A Field Effect Transistor (FET), as found in **Answer A**, is the right choice. The other choices are distractions to see if you read the question carefully.

T6B09 What are the names of the two electrodes of a diode?
A. Plus and minus
B. Source and drain
C. Anode and cathode
D. Gate and base

The diode has an anode and a cathode, so **Answer C** is the right choice. Plus and minus are battery terminals. You will find the other choices on various transistors and not diodes.

T6B10 Which of the following could be the primary gain-producing component in an RF power amplifier?
A. Transformer
B. Transistor
C. Reactor
D. Resistor

Transformers and resistors do not produce a circuit gain, so Answers A and D are incorrect. A reactor is not a radio circuit, so Answer C is also incorrect. The transistor in **Answer B** can be a gain-producing amplifier, so it is the correct option.

T6B11 What is the term that describes a transistor's ability to amplify a signal?
A. Gain
B. Forward resistance
C. Forward voltage drop
D. On resistance

Amplification implies that the output signal is larger than the input signal. Another name for this process is gain, so **Answer A** is correct. The others are good engineering terms, but they do not apply here.

6.5 T6C – Circuit Diagrams

6.5.1 Overview

The *Circuit Diagrams* question group in Subelement T6 introduces you to electrical circuit diagrams. The *Circuit Diagrams* group covers reading the symbols for circuit elements in electrical schematic diagrams. There is a total of 13 questions in this group of which one will be selected for the exam.

6.5.2 Questions

T6C01 What is the name of an electrical wiring diagram that uses standard component symbols?

A. Bill of materials
B. Connector pinout
C. Schematic
D. Flow chart

We saw examples of circuit schematic symbols in Figure 6.1. **Answer C** contains the correct choice. A bill of materials is a listing of things you bought. A flow chart is for software and not electrical circuits. A pinout is an important diagram, but it is not an all-inclusive diagram like the schematic, so this is not the best choice.

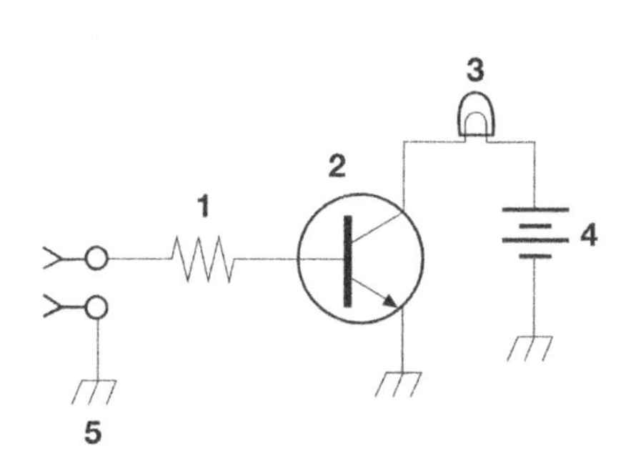

Figure 6.2: Figure T-1 for the Technician question pool.

T6C02 What is component 1 in figure T1?

A. Resistor
B. Transistor
C. Battery
D. Connector

Component 1 is a resistor, component 2 is a NPN transistor, component 3 is a lamp, and component 4 is a battery. **Answer A** is the correct choice.

T6C03 What is component 2 in figure T1?

A. Resistor
B. Transistor
C. Indicator lamp
D. Connector

Component 2 is the transistor, so **Answer B** is correct.

T6C04 What is component 3 in figure T1?
A. Resistor
B. Transistor
C. Lamp
D. Ground symbol

Component 3 is a lamp making **Answer C** the correct choice.

T6C05 What is component 4 in figure T1?
A. Resistor
B. Transistor
C. Battery
D. Ground symbol

Component 4 is a battery making **Answer C** the correct choice.

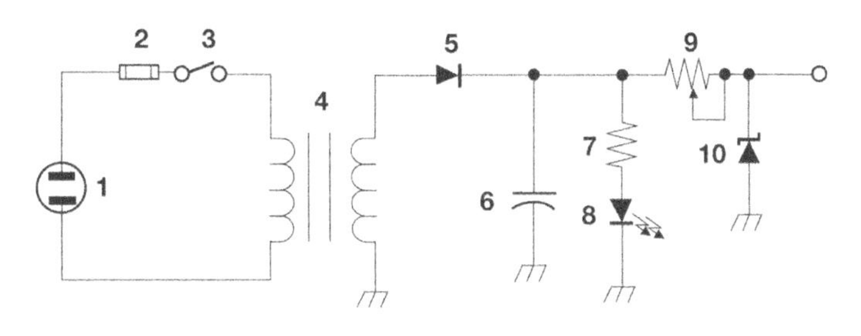

Figure 6.3: Figure T2 for the Technician question pool.

T6C06 What is component 6 in figure T2?
A. Resistor
B. Capacitor
C. Regulator IC
D. Transistor

Component 6 is a capacitor, as in **Answer B**. The resistor is component 7. There is no transistor or regulator IC in Figure T2.

T6C07 What is component 8 in figure T2?
A. Resistor
B. Inductor
C. Regulator IC
D. Light emitting diode

Component 8 is a Light Emitting Diode, as in **Answer D**.

T6C08 What is component 9 in figure T2?
A. Variable capacitor
B. Variable inductor
C. Variable resistor
D. Variable transformer

Component 9 is a resistor, in particular a variable resistor or a potentiometer, as in **Answer C**.

T6C09 What is component 4 in figure T2?
A. Variable inductor
B. Double-pole switch
C. Potentiometer
D. Transformer

Component 4 is a transformer, as in **Answer D**. Neither a double pole switch nor a variable inductor appears in Figure T2.

T6C10 What is component 3 in figure T3?
A. Connector
B. Meter
C. Variable capacitor
D. Variable inductor

Component 3 is a variable inductor, as in **Answer D**. Component 2 is a variable capacitor. Component 1 is a connector. Component 4 is an antenna. There is no meter in Figure T3.

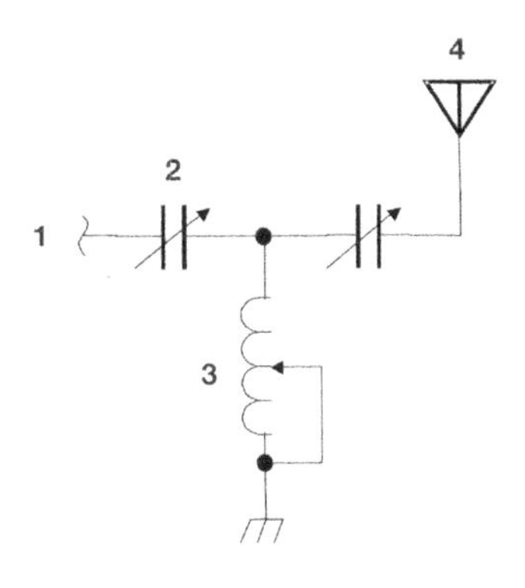

Figure 6.4: Figure T3 for the Technician question pool.

T6C11 What is component 4 in figure T3?

A. Antenna
B. Transmitter
C. Dummy load
D. Ground

Component 4 is an antenna, as in **Answer A**.

T6C12 What do the symbols on an electrical schematic represent?

A. Electrical components
B. Logic states
C. Digital codes
D. Traffic nodes

As you may suspect by now, the symbols on the schematic diagram represent electrical components, as in **Answer A**.

T6C13 Which of the following is accurately represented in electrical schematics?

A. Wire lengths
B. Physical appearance of components
C. The way components are interconnected
D. All of these choices are correct

The schematic diagram shows how the designer connected the components, as in **Answer C**. One cannot tell wiring length or the physical appearance of the components from the schematic.

6.6 T6D – Component Functions

6.6.1 Overview

The *Component Functions* question group in Subelement T6 goes into more depth about the properties of components found in radio circuits. The *Component Functions* group covers topics such as

- Rectification
- Switches
- Indicators
- Power supply components
- Resonant circuit
- Shielding
- Power transformers
- Integrated circuits

There is a total of 12 questions in this group of which one will be selected for the exam.

6.6.2 Questions

T6D01 Which of the following devices or circuits changes an alternating current into a varying direct current signal?

A. Transformer
B. Rectifier
C. Amplifier
D. Reflector

The transformer in Answer A converts a voltage from one level to another level. The rectifier in **Answer B** will convert AC current into a varying direct (not going in reverse) current, so this is the best choice to answer this question. The amplifier of Answer C will increase the magnitude of the signal, so this is not a correct choice. The reflector of Answer D is part of an antenna, so this is not related to the question.

T6D02 What is a relay?

A. An electrically-controlled switch
B. A current controlled amplifier
C. An optical sensor
D. A pass transistor

A relay has two major elements: a switch and an electromagnet that controls the switch setting. This corresponds to the choice in **Answer A**. The other devices do not act as switches, so they are distractions.

T6D03 What type of switch is represented by component 3 in figure T2?

A. Single-pole single-throw
B. Single-pole double-throw
C. Double-pole single-throw
D. Double-pole double-throw

Looking at both Figure 6.1 and Figure 6.3, component 3 is a single-pole, single-throw switch, so **Answer A** is the correct choice. Circuit schematic symbols for various other switches are in Figure 6.1. As you can see, the other choices do not match component 3.

T6D04 Which of the following displays an electrical quantity as a numeric value?

A. Potentiometer
B. Transistor
C. Meter
D. Relay

A potentiometer, a transistor, and a relay are all electronic components without a display. A meter usually has a display, so **Answer C** is the best choice among those given here.

T6D05 What type of circuit controls the amount of voltage from a power supply?
A. Regulator
B. Oscillator
C. Filter
D. Phase inverter

A regulator controls the amount of voltage from a power supply, which makes **Answer A** the correct choice. The other choices do not control power supplies.

T6D06 What component is commonly used to change 120V AC house current to a lower AC voltage for other uses?
A. Variable capacitor
B. Transformer
C. Transistor
D. Diode

A transformer is the usual component used to change AC sources from one voltage level to another. This makes **Answer B** the correct choice. Designers do not use capacitors, transistors, or diodes for this type of voltage change.

T6D07 Which of the following is commonly used as a visual indicator?
A. LED
B. FET
C. Zener diode
D. Bipolar transistor

The FET, the Zener diode, and the bipolar transistor do not emit visual light under normal conditions. However, manufacturers build the LED to emit light that designers can use it as a visual indicator, so **Answer A** is the correct choice.

T6D08 Which of the following is combined with an inductor to make a tuned circuit?
A. Resistor
B. Zener diode
C. Potentiometer
D. Capacitor

A tuned circuit, one that responds to a specific frequency, needs both a capacitor and an inductor, so **Answer D** is the best choice among those given here. The other components do not have the right properties by themselves to give a tuning action when paired with an inductor.

T6D09 What is the name of a device that combines several semiconductors and other components into one package?
A. Transducer
B. Multi-pole relay
C. Integrated circuit
D. Transformer

An integrated circuit "integrates" several semiconductor and other components into a single package, as in **Answer C**. The other choices are discrete components and not integrated circuits.

T6D10 What is the function of component 2 in Figure T1?
A. Give off light when current flows through it
B. Supply electrical energy
C. Control the flow of current
D. Convert electrical energy into radio waves

Looking at both Figure 6.1 and Figure 6.2, component 2 is a transistor. As we saw in earlier questions, designers use a transistor to control the flow of current, so **Answer C** is the right choice among those given here. The transistor controls the current by acting as a switch to turn the lamp (component 3) on and off.

T6D11 Which of the following is a resonant or tuned circuit?
A. An inductor and a capacitor connected in series or parallel to form a filter
B. A type of voltage regulator
C. A resistor circuit used for reducing standing wave ratio
D. A circuit designed to provide high fidelity audio

As we noted earlier, we need both an inductor and a capacitor for a tuned circuit, so you should be able to spot **Answer A** as the right choice. Regulators control voltage levels, and do not make resonant circuits. A purely resistive circuit reducing the Standing Wave Ratio (SWR) cannot produce resonance without additional reactive components. The audio circuit of Answer D is technically not a resonant circuit.

T6D12 Which of the following is a common reason to use shielded wire?
A. To decrease the resistance of DC power connections
B. To increase the current carrying capability of the wire
C. To prevent coupling of unwanted signals to or from the wire
D. To couple the wire to other signals

Shielded wired provides a ground return for stray signals, which is very helpful in reducing noise and other unwanted signals, so **Answer C** is the right choice. The shielding, technically, does none of the functions given in the other choices.

Chapter 7

T7 – STATION EQUIPMENT

7.1 Introduction

In this chapter, we will have a deeper dive into the characteristics of the station's Radio Frequency (RF) equipment that we began in Chapter 4. The questions cover both equipment in the amateur radio station and problems that the operator may encounter with their equipment and other communications equipment. This *Station Equipment* subelement has the following question groups:

A. Station equipment
B. Common transmitter and receiver problems
C. Antenna measurements and troubleshooting
D. Basic repair and testing

This will generate four questions on the Technician examination.

7.2 Radio Engineering Concepts

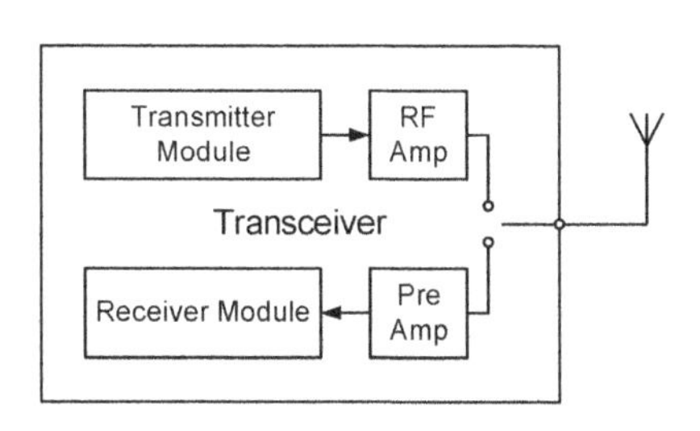

Figure 7.1: Amplifiers for the transmitter and receiver modules of a transceiver

Transceiver Elements The modern amateur radio Transceiver (XCVR) is a sophisticated piece of electronics with various signal processing elements inside the box to allow the operator to use it for both transmission and reception operations. In Figure 4.2 we saw some of the external user features. Questions in this chapter will cover interior circuits and performance characteristics.

Internal components in the transceiver that the questions cover are

Mixer — an electronic circuit found in transmitters and receivers that shifts a signal from one frequency region to another frequency region

Oscillator — an electronic circuit found in transmitters and receivers that generates a sinusoidal signal at a specified frequency

PTT Switch — the Push to Talk (PTT) switch is a circuit element that turns on the audio signal from the microphone into the transmitter when pressed, and cuts off the signal when released

Modulator — an electronic circuit in the transmitter that modifies a carrier based on the message signal

Amplifier — an electronic circuit to boost a signal; referring to Figure 7.1, a *pre-amplifier* boosts an input RF signal to improve reception, and an *RF amplifier* boosts a transmitter output signal before sending it to the antenna

There are several technical quality measures that we can apply to a transceiver. Quality measures that the questions cover include

Sensitivity — a measure of the ability of a receiver to detect an RF signal

Selectability — a measure of the ability of a receiver to discriminate between RF signals

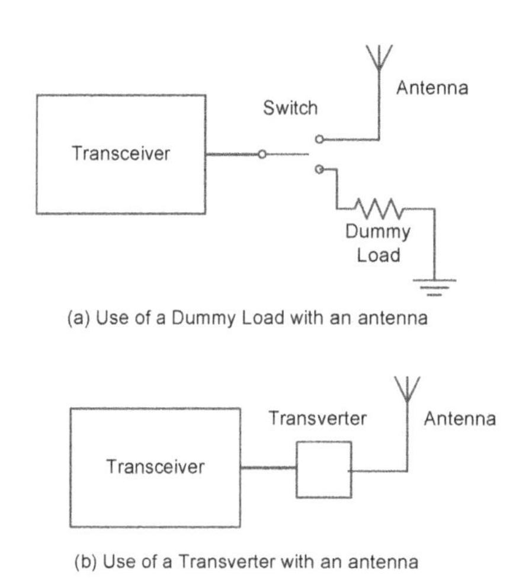

Figure 7.2: Interface between a transceiver and an antenna with a dummy load or a transverter

Antenna Interface The interface between the transceiver and the antenna is typically a single cable. However, there are times additional equipment is beneficial. The questions cover two interface options illustrated in Figure 7.2

Dummy Load — a resistor having the same impedance as the antenna that can handle the transceiver's RF power, and allows the user to make transceiver adjustments without transmitting over the air; the operator switches between the antenna and the load

Transverter — a device to shift the incoming or outgoing RF signal to match the transceiver's operating region, and allow a greater operating range; this is also known as either an *upconverter* or a *downconverter*

Antenna Feed Lines Your amateur station will use a feed line between the transceiver and the antenna. Usually this will be a co-axial cable, and not a normal hook-up wire used in electronics. Concerns with feed lines include

Feedline Integrity — the feedline's outer cladding and connections need to be intact because water in the feedline will cause problems; age, UV light, animals, and laying across hot roofs can compromise the cladding

Impedance Mismatch — a mismatch between the output impedance of the transmitter and the input impedance of the antenna will cause extra heating of the feedline; engineers use the Standing Wave Ratio (SWR) to characterize the mismatch with a SWR of 1:1 being a perfect match.

A SWR level of 2:1 is usually about the maximum mismatch the radio equipment will safely tolerate.

Electrical Measurements While operators can treat many aspects of modern amateur radio equipment as an "appliance" requiring little adjustment, there are times when you will need to make electrical measurements. Naturally, you will need the proper measurement equipment to do that job. Figure 7.3 illustrates four primary measurements:

Voltage Measurement is made with a *voltmeter* across the circuit element of interest; a special voltmeter is the Vacuum Tube Volt Meter (VTVM), which has a very high input resistance, and gives a high-quality measurement

Current Measurement is made with an *ammeter* measuring the current passing through the circuit element of interest

Resistance Measurement is made with an *ohmmeter* across the circuit element of interest; notice that there is not an energy source driving the circuit when making the measurement

SWR Measurement is made with a SWR meter that measures the power reflected from the antenna to the transmitter

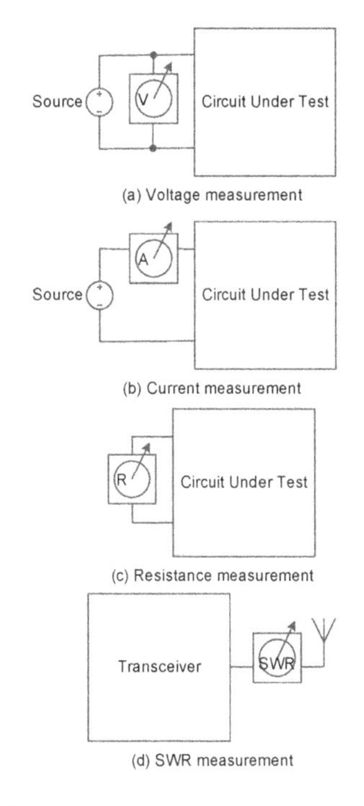

Figure 7.3: Metering configurations for making voltage, current, resistance, and SWR measurements.

Transceiver Problems While the transceiver elements are well designed, there are several problems that can occur due to the way someone operates it or from the environment. The questions cover two areas:

Over Deviation — a problem with Frequency Modulation (FM) transmitters when the input signal is too large, and causes a distorted transmission

Interference — Radio Frequency Interference (RFI) can be caused by either your equipment or by nearby communications equipment; examples are *fundamental*

overload where a strong carrier overwhelms the desired carrier, *harmonics* where an integer multiple of a carrier is produced by a transmitter, and a *spurious emission* where a transmitter produces a signal outside the normal, desired transmission bandwidth.

Operators resort to filters of various types and ferrite chokes to block these signals, depending upon their source.

Consumer Electronics Problems The amateur radio station components are not the only devices in your home or radio shack that may be emitting or receiving RF signals. The existence of these problems may not indicate a problem with your station's equipment, but may be a problem with the consumer electronics. In particular, the amateur radio operator needs to be aware of

Part 15 Devices — these are low-power, unlicensed consumer devices, such as Wi-Fi or Bluetooth devices, which share parts of the radio spectrum with the Amateur Radio Service; this is a potential place where mutual interference is possible, and users must work out accommodations

Telephones — amateur radio signals can enter telephones and cause interference; a filter may be necessary to remove the station's signal at the telephone

Televisions — amateur radio signals can enter television receivers and cause interference; a filter may be necessary to remove the station's signal at the television

If a neighbor does report a problem to you, be sure that your station is operating properly, and that you have properly connected all your RF cables before resorting to adding filters to anyone's equipment.

7.3 T7A – Station Equipment

7.3.1 Overview

The *Station Equipment* question group in Subelement T7 brings in more details about the equipment for a radio shack. The *Station Equipment* group covers topics such as

- Receivers
- Transmitters
- Transceivers
- Transverters
- Transmit and receive amplifiers

There is a total of 11 questions in this group of which one will be selected for the exam.

7.3.2 Questions

T7A01 Which term describes the ability of a receiver to detect the presence of a signal?

A. Linearity
B. Sensitivity
C. Selectivity
D. Total Harmonic Distortion

Linearity is an operation that mathematics describes by a straight line, selectivity is the ability to distinguish between signals, and total harmonic distortion is a measure of signal quality. None of these describes the ability to detect a signal. Sensitivity, as in **Answer B**, is this detection ability.

T7A02 What is a transceiver?

A. A type of antenna switch
B. A unit combining the functions of a transmitter and a receiver
C. A component in a repeater which filters out unwanted interference
D. A type of antenna matching network

The word Transceiver (XCVR) is a combination of "transmitter" and "receiver," which makes **Answer B** the right choice. An antenna switch allows the operator to choose between multiple antennas. A XCVR is not a filter, and it is not a special circuit for impedance matching, which makes the other choices all incorrect.

T7A03 Which of the following is used to convert a radio signal from one frequency to another?

A. Phase splitter
B. Mixer
C. Inverter
D. Amplifier

A mixer converts a signal from one frequency to another, so **Answer B** is the right choice. A phase splitter separates signals into multiple phases or polarities, an inverter changes a signal's polarity, and amplifier increases a signal's amplitude, so these are incorrect choices.

T7A04 Which term describes the ability of a receiver to discriminate between multiple signals?

A. Discrimination ratio
B. Sensitivity
C. Selectivity
D. Harmonic Distortion

Based on the earlier question, did you spot selectivity in **Answer C** as the correct choice for the ability of a receiver to discriminate between signals?

T7A05 What is the name of a circuit that generates a signal at a specific frequency?
A. Reactance modulator
B. Product detector
C. Low-pass filter
D. Oscillator

The implication here is that the signal should be a sinusoidal-shaped signal. An oscillator produces that type of signal, so **Answer D** is the right choice. A reactance modulator is a form of FM transmitter, a product detector demodulates both Continuous Wave (CW) and Single Sideband (SSB) signals in a receiver, and a low-pass filter removes components from a signal, so these are all incorrect.

T7A06 What device converts the RF input and output of a transceiver to another band?
A. High-pass filter
B. Low-pass filter
C. Transverter
D. Phase converter

Filters keep the input and output signals in the same frequency band, so Answers A and B are incorrect. A phase converter does not change frequency either. The transverter of **Answer C**, a device that moves a signal from one frequency to another, is the correct choice.

T7A07 What is meant by term "PTT"?
A. Pre-transmission tuning to reduce transmitter harmonic emission
B. Precise tone transmissions used to limit repeater access to only certain signals
C. A primary transformer tuner use to match antennas
D. The push to talk function which switches between receive and transmit

The Push to Talk (PTT) function switches between receive and transmit, so **Answer D** is the right choice. The others are to distract you.

T7A08 Which of the following describes combining speech with an RF carrier signal?
A. Impedance matching
B. Oscillation
C. Modulation
D. Low-pass filtering

Impedance matching assures devices have maximum power transfer, oscillation is a time variable signal, and low-pass filtering removes components from a signal, so none of these perform the operation described in the question. Modulation, as in **Answer C**, is the combining of a message signal with a carrier, so this is the right choice.

T7A09 What is the function of the SSB/CW-FM switch on a VHF power amplifier?

A. Change the mode of the transmitted signal
B. Set the amplifier for proper operation in the selected mode
C. Change the frequency range of the amplifier to operate in the proper portion of the band
D. Reduce the received signal noise

Amplifiers usually have an output filter to prevent transmitting unwanted emission signals. This SSB/CW-FM switch has two settings: one for narrow-band transmission (SSB/CW) and one for wide-band transmission (FM). This makes **Answer B** the correct choice. The transmitter and not the amplifier changes the mode or the operating frequency, so Answers A and C are incorrect. This is a transmission amplifier and not a reception amplifier, so Answer D is incorrect in this application.

T7A10 What device increases the low-power output from a handheld transceiver?

A. A voltage divider
B. An RF power amplifier
C. An impedance network
D. All of these choices are correct

A voltage divider will split the signal, so this is not a good choice. Circuit designers use an impedance network in matching the transceiver with the antenna, so this will not increase the low-power signal. Since these two are incorrect choices, Answer D is also incorrect. The best choice of those given here is the RF amplifier of **Answer B**.

T7A11 Where is an RF preamplifier installed?

A. Between the antenna and receiver
B. At the output of the transmitter's power amplifier
C. Between a transmitter and antenna tuner
D. At the receiver's audio output

Operators use a preamplifier to increase the signal going into a receiver, so **Answer A** is the right choice. Each of the other suggested locations is improper for a preamplifier.

7.4 T7B – Common Transmitter and Receiver Problems

7.4.1 Overview

The *Common Transmitter and Receiver Problems* question group in Subelement T7 involves issues operators face with their equipment. The *Common Transmitter and Receiver Problems* group covers topics such as

- Symptoms of overload and overdrive
- Distortion
- Causes of interference
- Interference and consumer electronics
- Part 15 devices
- Over-modulation
- RF feedback
- Off frequency signals

There is a total of 12 questions in this group of which one will be selected for the exam.

7.4.2 Questions

T7B01 What can you do if you are told your FM handheld or mobile transceiver is over-deviating?

A. Talk louder into the microphone
B. Let the transceiver cool off
C. Change to a higher power level
D. Talk farther away from the microphone

Speaking louder into the FM transceiver will increase, and not lower, the over deviation, so Answer A will not fix the problem. Answer B will not cure the problem because a too-strong an audio input, and not heat, causes over deviation, so this is a bad choice. Answer C will also not help since it does not reduce audio input intensity. **Answer D** is the right choice because it will decrease the audio input signal's amplitude and reduce the deviations.

T7B02 What would cause a broadcast AM or FM radio to receive an amateur radio transmission unintentionally?

A. The receiver is unable to reject strong signals outside the AM or FM band
B. The microphone gain of the transmitter is turned up too high
C. The audio amplifier of the transmitter is overloaded
D. The deviation of an FM transmitter is set too low

In this question, we are assuming that the amateur station is operating in proper order, and the control operator is using it properly. In that case, there will be no microphone gain issues, and the signal does not overload the amplifiers. Low FM deviation will make the signal even more confined on the amateur bands. In this case, the problem is with the broadcast receiver, and it is not functioning properly, as in **Answer A**, so this is the right choice.

T7B03 Which of the following can cause radio frequency interference?
A. Fundamental overload
B. Harmonics
C. Spurious emissions
D. All of these choices are correct

Each of the choices given in Answers A, B, and C may cause Radio Frequency Interference (RFI). Therefore, the best choice to answer this question is **Answer D**.

T7B04 Which of the following is a way to reduce or eliminate interference from an amateur transmitter to a nearby telephone?
A. Put a filter on the amateur transmitter
B. Reduce the microphone gain
C. Reduce the SWR on the transmitter transmission line
D. Put a RF filter on the telephone

If the operator's equipment is in good working order, the solutions proposed in Answers A and B will not solve the problem. However, given the information in the question, the best choice is the RF filter of **Answer D**. Reducing the SWR, as in Answer C, will not remove spurious signals leaking into the telephone system.

T7B05 How can overload of a non-amateur radio or TV receiver by an amateur signal be reduced or eliminated?
A. Block the amateur signal with a filter at the antenna input of the affected receiver
B. Block the interfering signal with a filter on the amateur transmitter
C. Switch the transmitter from FM to SSB
D. Switch the transmitter to a narrow-band mode

Switching the modulation mode from FM to SSB or changing from wideband to narrow band mode will not prevent the amateur station's signal from entering the non-amateur radio equipment, so these are not good choices. Blocking the signal at the amateur station's antenna will work, but it may prevent legitimate amateur communications from happening, so this is not the best choice. Blocking the signal at the non-amateur receiver is the best solution, so **Answer A** is the right choice.

T7B06 Which of the following actions should you take if a neighbor tells you that your station's transmissions are interfering with their radio or TV reception?

A. Make sure that your station is functioning properly and that it does not cause interference to your own radio or television when it is tuned to the same channel
B. Immediately turn off your transmitter and contact the nearest FCC office for assistance
C. Tell them that your license gives you the right to transmit and nothing can be done to reduce the interference
D. Install a harmonic doubler on the output of your transmitter and tune it until the interference is eliminated

Answers C and D are bad amateur practice, so these are not good choices to answer this question. Answer B is a bad choice because the Federal Communications Commission (FCC) office will not be able to assist you. **Answer A** gives the correct procedure, so that is the correct answer.

T7B07 Which of the following can reduce overload to a VHF transceiver from a nearby FM broadcast station?

A. RF preamplifier
B. Double-shielded coaxial cable
C. Using headphones instead of the speaker
D. Band-reject filter

The RF preamplifier in Answer A will make the interfering signal stronger, so this is not a good choice. The shielded co-axial cable will conduct the signal well, but not remove the interference, so this is not a good choice. The headphones in Answer C are a silly distraction. The band-reject filter in **Answer D** will "notch-out" the interfering signal making this the correct option here.

T7B08 What should you do if something in a neighbor's home is causing harmful interference to your amateur station?

A. Work with your neighbor to identify the offending device
B. Politely inform your neighbor about the rules that prohibit the use of devices which cause interference
C. Check your station and make sure it meets the standards of good amateur practice
D. All of these choices are correct

Technically, each of the steps in Answers A, B, and C are proper for resolving this issue. Therefore, **Answer D** is the best choice to answer this question.

T7B09 What is a Part 15 device?

A. An unlicensed device that may emit low powered radio signals on frequencies used by a licensed service
B. An amplifier that has been type-certified for amateur radio
C. A device for long distance communications using special codes sanctioned by the International Amateur Radio Union
D. A type of test set used to determine whether a transmitter is in compliance with FCC regulation 91.15

Part 15 devices are low-power, short range, unlicensed devices that must share the radio spectrum with other users in a non-interfering manner, as described in **Answer A**. They are not amateur radio devices, nor are they used for long-range communications. Answer D is a silly distractor.

T7B10 What might be a problem if you receive a report that your audio signal through the repeater is distorted or unintelligible?

A. Your transmitter may be slightly off frequency
B. Your batteries may be running low
C. You could be in a bad location
D. All of these choices are correct

Technically, each of the statements in Answers A, B, and C might be a cause of a bad signal. Therefore, **Answer D** is the best choice to answer this question.

T7B11 What is a symptom of RF feedback in a transmitter or transceiver?

A. Excessive SWR at the antenna connection
B. The transmitter will not stay on the desired frequency
C. Reports of garbled, distorted, or unintelligible transmissions
D. Frequent blowing of power supply fuses

The feedback described in the question does not affect antenna SWR, transmitter tuning, or power supplies. However, it will distort the transmitted signal, which makes **Answer C** the correct choice for this question.

T7B12 What should be the first step to resolve cable TV interference from your ham radio transmission?

A. Add a low pass filter to the TV antenna input
B. Add a high pass filter to the TV antenna input
C. Add a preamplifier to the TV antenna input
D. Be sure all TV coaxial connectors are installed properly

The simplest solution is the first step to try to solve the problem. The first step is to be sure all the connectors are properly installed, as in **Answer D**. Adding filters would only be after completing this first step, so they are not the best choice. Adding an amplifier may make the situation worse, so this is not a good choice.

7.5 T7C – Antenna Measurements and Troubleshooting

7.5.1 Overview

The *Antenna Measurements and Troubleshooting* question group in Subelement T7 introduces you to antenna concepts for the amateur station. The *Antenna Measurements and Troubleshooting* group covers topics such as

- Measuring SWR
- Dummy loads
- Coaxial cables
- Feed line failure modes

There is a total of 12 questions in this group of which one will be selected for the exam.

7.5.2 Questions

T7C01 What is the primary purpose of a dummy load?

A. To prevent transmitting signals over the air when making tests
B. To prevent over-modulation of your transmitter
C. To improve the radiation from your antenna
D. To improve the signal to noise ratio of your receiver

A dummy load (see Figure 4.2) is a resistor that can absorb the transmission from your rig and keep it from going out over the air while you are doing maintenance or tuning activities. This makes **Answer A** the correct choice. Answers B and C cannot be cured by this device, so they are not good choices. Answer D is technically not true.

T7C02 Which of the following instruments can be used to determine if an antenna is resonant at the desired operating frequency?

A. A VTVM
B. An antenna analyzer
C. A Q meter
D. A frequency counter

The Vacuum Tube Volt Meter (VTVM) and the frequency counter do not determine resonance. Operators use the "Q" meter to measure the resonance of electronic circuits, and not antennas, so these are not good choices. The manufacturer designed the antenna analyzer to measure antenna characteristics making **Answer B** the correct choice.

T7C03 What, in general terms, is standing wave ratio (SWR)?

A. A measure of how well a load is matched to a transmission line
B. The ratio of high to low impedance in a feed line
C. The transmitter efficiency ratio
D. An indication of the quality of your station's ground connection

The SWR measures how well the load matches the transmitter (or, how well any two items match). This is a measure of how close they are to having the same impedance, which makes **Answer A** the right choice. Answers B, C, and D are good things to know, but they are not SWR.

T7C04 What reading on an SWR meter indicates a perfect impedance match between the antenna and the feed line?

A. 2 to 1
B. 1 to 3
C. 1 to 1
D. 10 to 1

A perfect match is 1:1, as in **Answer C**. The others represent differing degrees of mismatch, so they are incorrect.

T7C05 Why do most solid-state amateur radio transmitters reduce output power as SWR increases?

A. To protect the output amplifier transistors
B. To comply with FCC rules on spectral purity
C. Because power supplies cannot supply enough current at high SWR
D. To improve the impedance match to the feed line

Answer A has the correct reasoning because supplying the power under a high SWR condition may damage the amplifier transistors. There is no corresponding FCC spectral purity rule, so Answer B is incorrect. The SWR does not affect the power supply, which makes Answer C incorrect. Decreasing power will not change the SWR making Answer D incorrect too.

T7C06 What does an SWR reading of 4:1 indicate?

A. Loss of -4dB
B. Good impedance match
C. Gain of +4dB
D. Impedance mismatch

Device gain and loss, as in Answers A and C, do not directly have anything to do with the impedance match, so these options are incorrect. As we saw above, a 1:1 is a good match, so this is not true here. An impedance mismatch, as in **Answer D**, is the correct choice.

T7C07 What happens to power lost in a feed line?
A. It increases the SWR
B. It comes back into your transmitter and could cause damage
C. It is converted into heat
D. It can cause distortion of your signal

Electronic circuit elements, not just feedlines, frequently convert power losses to heat, as in **Answer C**. It does not increase the SWR or damage the transmitter. Nor does it distort the signal.

T7C08 What instrument other than an SWR meter could you use to determine if a feed line and antenna are properly matched?
A. Voltmeter
B. Ohmmeter
C. Iambic pentameter
D. Directional wattmeter

A voltmeter only measures signal voltages, and not power, in this sense, so Answer A is not a good choice. The ohmmeter can give an indication of resistance, but most voltmeters do not allow you to specify an operating frequency, so it will not help in SWR measurements. Answer C is for the English majors, and not the radio engineers. The directional wattmeter will tell transmitted and reflected power, which are the measurements needed for SWR computation, so **Answer D** is the right choice.

T7C09 Which of the following is the most common cause for failure of coaxial cables?
A. Moisture contamination
B. Gamma rays
C. The velocity factor exceeds 1.0
D. Overloading

If moisture gets into the co-axial cable, it can cause the cable to fail, so **Answer A** is the correct choice. "Gamma rays" is a silly distraction. Having a velocity factor larger than 1 implies that the cable transmits electromagnetic radiation faster than the speed of light, which is not a reasonable answer. Overloading can lead to damage in the co-ax. However, this is beyond the legal power limits for amateur stations, so it will not be a common failure mode for amateur stations.

T7C10 Why should the outer jacket of coaxial cable be resistant to ultraviolet light?
A. Ultraviolet resistant jackets prevent harmonic radiation
B. Ultraviolet light can increase losses in the cable's jacket
C. Ultraviolet and RF signals can mix together, causing interference
D. Ultraviolet light can damage the jacket and allow water to enter the cable

Each of the reasons given in Answers A, B, and C is engineering technobabble.

Preventing damaging the jacket and letting water in, as in **Answer D**, is the right choice.

T7C11 What is a disadvantage of air core coaxial cable when compared to foam or solid dielectric types?

A. It has more loss per foot
B. It cannot be used for VHF or UHF antennas
C. It requires special techniques to prevent water absorption
D. It cannot be used at below freezing temperatures

The reasons given in Answers A, B, and D are all electronically incorrect. Using air core coaxial cable requires that the user avoids water absorption, so **Answer C** is the correct choice.

T7C12 What does a dummy load consist of?

A. A high-gain amplifier and a TR switch
B. A non-inductive resistor and a heat sink
C. A low voltage power supply and a DC relay
D. A 50 ohm reactance used to terminate a transmission line

Here we are back to the dummy load again. Earlier we said it was a resistor, and now we have a bit more detail. It is not any old resistor, but a non-inductive one with a heat sink, as in **Answer B**. Many dummy loads are $50\,\Omega$, but they are not reactances, so Answer D is incorrect. Answers A and C are silly distractors.

7.6 T7D – Basic Repair and Testing

7.6.1 Overview

The *Basic Repair and Testing* question group in Subelement T7 introduces you to electrical measurement and construction techniques found in radio circuits. The *Basic Repair and Testing* group covers topics such as

- Soldering
- Using basic test instruments
- Connecting a voltmeter, ammeter, or ohmmeter

There is a total of 12 questions in this group of which one will be selected for the exam.

7.6.2 Questions

T7D01 Which instrument would you use to measure electric potential or electromotive force?
A. An ammeter
B. A voltmeter
C. A wavemeter
D. An ohmmeter

An ammeter measures current, a wavemeter measures RF energy, and an ohmmeter measures electrical resistance. **Answer B** gives the correct response, which is a voltmeter.

T7D02 What is the correct way to connect a voltmeter to a circuit?
A. In series with the circuit
B. In parallel with the circuit
C. In quadrature with the circuit
D. In phase with the circuit

Users place meters either in series or in parallel with the device under test, so the right answer will come from Answers A or B. Answers C and D are incorrect because they do not deal with series and parallel. Voltmeters measure in parallel (voltage across an element), while ammeters measure in series (current through an element). The correct answer here is **Answer B** since we are measuring voltage, and not measuring current.

T7D03 How is an ammeter connected to a circuit?
A. In series with the circuit
B. In parallel with the circuit
C. In quadrature with the circuit
D. In phase with the circuit

This question is a complement to the previous question. Answers C and D are incorrect because they do not deal with series and parallel. Voltmeters measure in parallel, while ammeters measure in series. The correct answer here is **Answer A** since we are measuring current, and not measuring voltage.

T7D04 Which instrument is used to measure electric current?
A. An ohmmeter
B. A wavemeter
C. A voltmeter
D. An ammeter

As you may be able to guess, an ammeter, as in **Answer D**, measures current in amperes. The ohmmeter measures resistance. The wavemeter measures RF energy.

The voltmeter measures electrical potential.

T7D05 What instrument is used to measure resistance?

A. An oscilloscope
B. A spectrum analyzer
C. A noise bridge
D. An ohmmeter

An oscilloscope is for measuring time-varying waveforms, but it is not for measuring resistance. A spectrum analyzer will show the signal in the frequency domain, but not resistance, so this is also incorrect. The noise bridge will tell you if the circuit is properly tuned, but not measure resistance. The ohmmeter of **Answer D** is the correct choice.

T7D06 Which of the following might damage a multimeter?

A. Measuring a voltage too small for the chosen scale
B. Leaving the meter in the milliamps position overnight
C. Attempting to measure voltage when using the resistance setting
D. Not allowing it to warm up properly

Small signals will not cause much needle swing, so it should not damage the meter making this a bad choice. Leaving the meter on overnight might run down the internal battery, but not cause lasting damage, so this is also a bad choice. Improperly using the settings could cause damage, so **Answer C** is the best choice among those given. Warm up is not a problem with modern meters.

T7D07 Which of the following measurements are commonly made using a multimeter?

A. SWR and RF power
B. Signal strength and noise
C. Impedance and reactance
D. Voltage and resistance

As indicated in **Answer D**, a multimeter can usually measure voltage and resistance. You will not find the other combinations on typical multimeters.

T7D08 Which of the following types of solder is best for radio and electronic use?

A. Acid-core solder
B. Silver solder
C. Rosin-core solder
D. Aluminum solder

One typically uses acid-core solder for oxidized metals and silver solder for jewelry, but not often in electronics. Aluminum solder is for soldering aluminum or brass, and not electronic components. Resin-core solder, as in **Answer C**, is the right choice among those given.

T7D09 What is the characteristic appearance of a cold solder joint?
  A. Dark black spots
  B. A bright or shiny surface
  C. A grainy or dull surface
  D. A greenish tint

A "cold" solder joint, one that does not make good electrical contact between the components, usually has a dull surface, as in **Answer C**. A bright shiny surface is the one we want to see, so that is the wrong answer here. Answers A and D are to distract you.

T7D10 What is probably happening when an ohmmeter, connected across an unpowered circuit, initially indicates a low resistance and then shows increasing resistance with time?
  A. The ohmmeter is defective
  B. The circuit contains a large capacitor
  C. The circuit contains a large inductor
  D. The circuit is a relaxation oscillator

The act of making the measurement provides a small activation energy to the circuit that makes the circuit operation sound like a capacitor is charging up, so **Answer B** is the right choice.

T7D11 Which of the following precautions should be taken when measuring circuit resistance with an ohmmeter?
  A. Ensure that the applied voltages are correct
  B. Ensure that the circuit is not powered
  C. Ensure that the circuit is grounded
  D. Ensure that the circuit is operating at the correct frequency

If you measure resistance when the circuit is powered, that is energized, you are engaged in a dangerous activity. This makes **Answer B** the correct choice. Answers A and D indicate that you have energized the circuit, so they are bad choices. Having the circuit grounded will not help with the resistance measurement.

T7D12 Which of the following precautions should be taken when measuring high voltages with a voltmeter?
  A. Ensure that the voltmeter has very low impedance
  B. Ensure that the voltmeter and leads are rated for use at the voltages to be measured
  C. Ensure that the circuit is grounded through the voltmeter
  D. Ensure that the voltmeter is set to the correct frequency

From a safety point of view, having the voltmeter and leads rated for the measurement range is an important precaution, so **Answer B** is the right choice. Voltmeter impedances are normally very high, so Answer A is technically incorrect. Grounding the circuit through the volt meter could be a safety problem. Volt meters normally do not have frequency settings.

Chapter 8

T8 – MODULATION MODES

8.1 Introduction

We had an introduction to radio modulation in Chapter 2 where we saw the definition of the carrier, the types of analog modulation, and the associated bandwidth. In this chapter, we will see questions that build upon that knowledge, and then apply it to amateur radio activities. This *Modulation Modes* subelement has the following question groups:

A. Modulation modes
B. Amateur satellite operation
C. Operating activities
D. Non-voice and digital communications

This will generate four questions on the Technician examination.

8.2 Radio Engineering Concepts

In this chapter, we will see more details on signal modulation than we introduced earlier. We first start with some review before we introduce new concepts needed for the questions. We will break the modulation into two classes

Analog Modulation — the message source is a continuous signal, and the carrier makes continuous changes

Digital Modulation — the message source is a discrete signal (like a bit), and the carrier makes discrete, step-like changes

Traditionally, the transceiver only operated with analog modulation modes, and operators used the computer to generate the digital modulation with either a modem or a sound card. Newer transceivers allow the user to directly plug the computer into the transceiver because they are "digital ready."

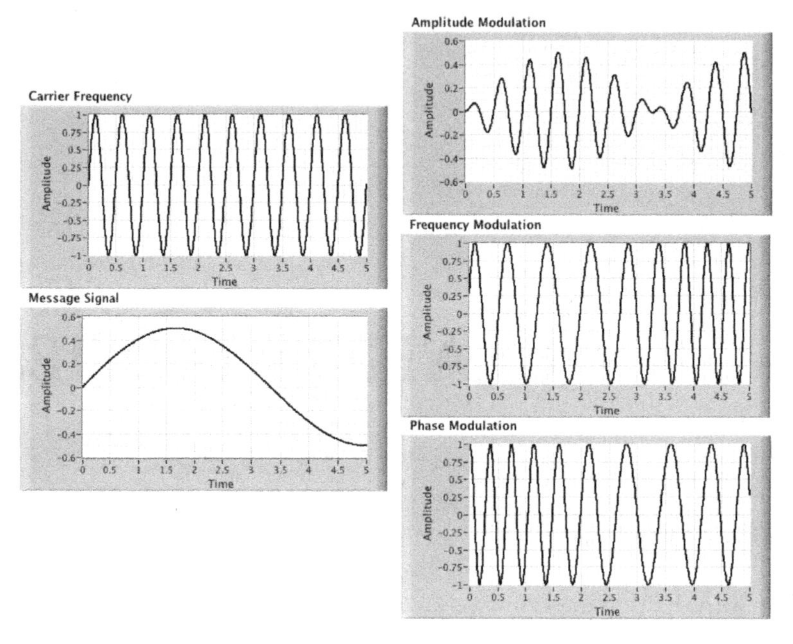

Figure 8.1: Example of analog amplitude, frequency, and phase modulation for an analog message signal.

Analog Modulation As we saw in Chapter 2, there are two major types of analog modulation: amplitude modulation and angle modulation. Amplitude Modulation (AM) uses the message signal to modify the carrier's amplitude. Angle modulation uses the message signal to modify either the carrier's frequency or phase angle. Figure 8.1 illustrates a carrier and a message signal on the left-hand side of the figure. The right-hand side illustrates amplitude, frequency, and phase modulation of the carrier by that message. AM comes in two major classes

Single Sideband — uses one copy of the message signal's frequency domain content in the transmission

Dual Sideband — uses both copies of the message signal's frequency domain content in the transmission

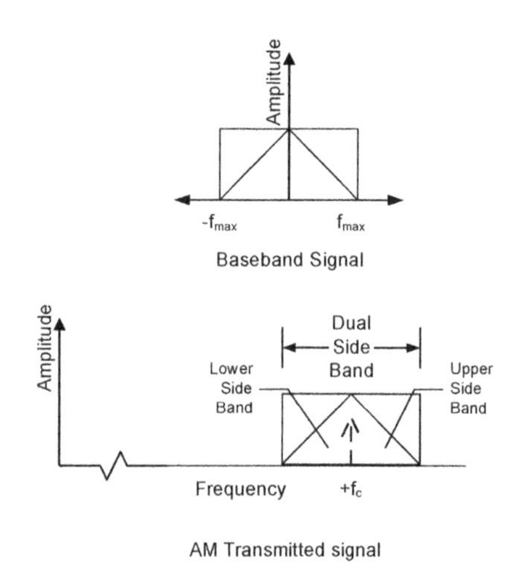

Figure 8.2: Amplitude modulation spectrum for DSB and SSB relative to the carrier f_c.

Single Sideband (SSB) comes in two modes: Upper Side Band (USB) and Lower Side Band (LSB). Dual Sideband (DSB) also comes in two modes: Dual Sideband - Residual Carrier (DSB-RC) where the transmitter sends an unmodulated copy of the carrier to aid in reception, and Dual Sideband - Suppressed Carrier (DSB-SC) where the transmitter does not send a copy of the carrier. DSB-RC is the format used in commercial broadcast AM transmissions. Figure 8.2 shows the difference between SSB and DSB in the frequency domain. The *baseband* block is the message signal as seen by a spectrum analyzer. When the message signal amplitude modulates the carrier, a copy of the baseband spectrum appears at the carrier location, f_c, in frequency space. DSB transmits the full copy of the spectrum. LSB transmits only the half of the spectrum below the carrier, while USB transmits only the half of the spectrum above the carrier. Neither SSB mode transmits a copy of the carrier. Both sidebands contain all the necessary information to recover the message signal in the receiver. SSB uses less transmission bandwidth than DSB at the expense of a bit more complicated electronics than DSB.

Angle Modulation also comes in two major classes illustrated in Figure 8.1

Frequency Modulation — uses the message signal to modify the carrier's frequency during the transmission

Phase Modulation — uses the message signal to modify the carrier's phase during the transmission

Amateur radio operators generally use Frequency Modulation (FM) over Phase Modulation (PM) by convention. We will see how digital modes use both approaches. Notice that in FM and PM parts of Figure 8.1, the carrier total phase varies with the message signal, while the amplitude stays constant.

Digital Modulation We build on the analog modulation concepts to develop digital modulation. Digital modulation does not use AM for transmission, but forms of angle modulation. Because the message signal in digital modulation only takes on discrete values, the frequency or phase only takes discrete values as well. Engineers call this *shift keying*. The major classes of digital modulation are

Frequency Shift Keying — uses the message signal to step the carrier's frequency during the transmission

Phase Shift Keying — uses the message signal to step the carrier's phase during the transmission

Figure 8.3 illustrates a digital message signal modifying a carrier with Frequency Shift Keying (FSK) and Phase Shift Keying (PSK). The top part of the figure shows the digital data. The middle of the figure shows the PSK waveform, while the bottom of the figure shows the FSK waveform. Notice that the phase and frequency change the waveform in step with the digital state changes.

Spread Spectrum Modulation Spread Spectrum (SS) modulation is a specific form of digital modulation. Your digital cell phone may use a form of this modulation. Its main strength is that it allows multiple users to transmit on the same carrier frequency without causing noticeable interference between them. There are two

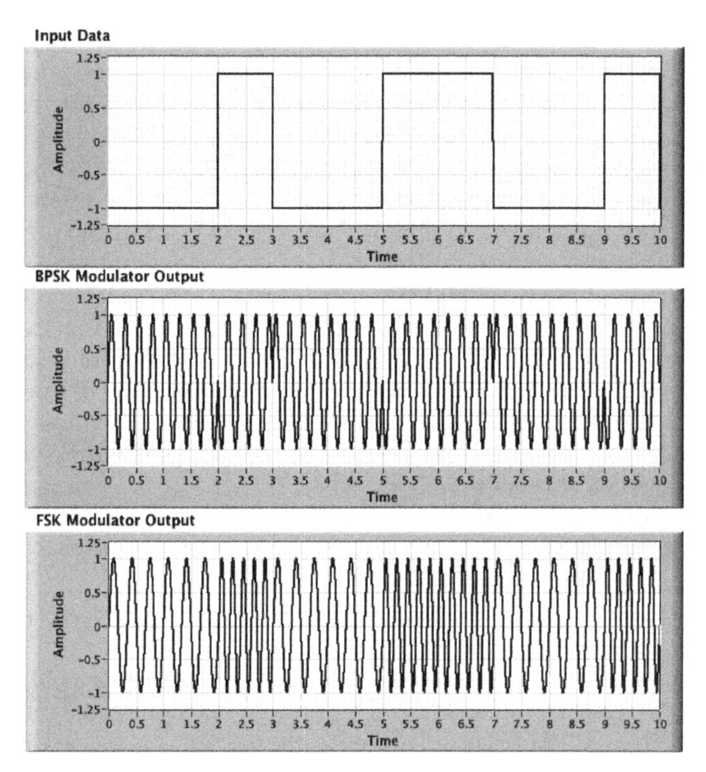

Figure 8.3: Example of digital phase and frequency modulation for a data message signal

methods for generating the SS signal

Direct Sequence Spread Spectrum — the use of a high-speed random digital signal to combine with the baseband message signal to produce a very high speed random-looking sequence; systems typically transmit with PSK modulation

Frequency Hopping Spread Spectrum — the use of a high-speed random offset to change the carrier frequency, so it hops rapidly and for a short duration across the frequency band

When we say "random" in these definitions, it is a long sequence of random-looking bits whose sequence is known to the transmitter and the receiver, so the receiver can decode the message properly. The Direct Sequence Spread Spectrum (DSSS) is the type used in your cell phone. The Frequency Hopping Spread Spectrum (FHSS) is used more in military communications.

Modulation Bandwidth The amount of frequency space that we need to transmit a signal depends on two things: the underlying characteristics of the signal, and the modulation mode we use in sending. In mathematical theory, all transmissions require an infinite bandwidth to send without distortion. However, engineers have

Table 8.1: Baseband and Modulated Bandwidth Estimates

Baseband Signal	Baseband Bandwidth
CW	150 Hz
Voice	2 kHz to 3 kHz
Digital Signals	Approx. the data transmission rate (bits/second)
Slow-scan TV	3 kHz
Fast-Scan TV	6 MHz
Modulated Type	**Modulated Signal Bandwidth**
Single Sideband Voice	2 kHz to 3 kHz
Dual Sideband Voice	6 kHz
Narrowband FM Voice	15 kHz
Commercial Broadcast FM	75 kHz

found that by using filters and clever signal processing, we can use much less bandwidth if we are willing to tolerate less than perfect conditions. This is usually good enough for most people. Table 8.1 lists common analog bandwidth estimates for transmitting various signals both at *baseband* (unmodulated), where they are formed, and as modulated signals. For digital bandwidth estimates, the bandwidth is approximately 1 Hz per bit per second of digital data rate.

8.3 T8A – Modulation Modes

8.3.1 Overview

The *Modulation Modes* question group in Subelement T8 introduces you to standard Radio Frequency (RF) modulation methods found in Amateur radio. The *Modulation Modes* group covers topics such as

- Modulation modes
- Bandwidth of various signals
- Choice of emission type

There is a total of 11 questions in this group of which one will be selected for the exam.

8.3.2 Questions

T8A01 Which of the following is a form of amplitude modulation?

A. Spread-spectrum
B. Packet radio
C. Single sideband
D. Phase shift keying (PSK)

Amplitude modulation means changing the carrier amplitude in response to the input message signal. Amplitude modulation is a form of analog signal modulation. The spread spectrum modulation of Answer A is a form a digital modulation usually executed by changing either the phase or the frequency of the carrier, so it is not AM. The packet radio choice of Answer B is a baseband format, and not a modulation format (one can send it with either analog or digital modulation), so this is also incorrect. The singe sideband of **Answer C** is a well-known form of AM, and this is the right choice. The PSK of Answer D is a form of digital modulation, so it is not a correct choice here.

T8A02 What type of modulation is most commonly used for VHF packet radio transmissions?

A. FM
B. SSB
C. AM
D. PSK

Current amateur practice uses a form of FM for Very High Frequency (VHF) packet radio, so **Answer A** is the correct choice. The SSB and AM of Answers B and C are fine for voice, but do not perform well with digital signals, so these are incorrect choices. The PSK of Answer D is technically possible, but it is not currently part of amateur practice, so this is also incorrect.

T8A03 Which type of voice mode is most often used for long-distance (weak signal) contacts on the VHF and UHF bands?

A. FM
B. DRM
C. SSB
D. PM

While most of the choices here are common amateur radio transmission options, the question is giving us a hint of the specific mode the question designers are looking for. To successfully receive a signal under these conditions, the receiving station desires to use as narrow a receiver bandwidth as possible to reduce the background noise under these conditions. This means that SSB in **Answer C** is the best choice because it has the narrowest bandwidth. The FM and PM of Answers A and D have much wider bandwidths than SSB, so reception might not be successful. The Digital Radio Mondial (DRM) of Answer B is a commercial broadcast standard, and not part of the Amateur Radio Service, so it is incorrect.

T8A04 Which type of modulation is most commonly used for VHF and UHF voice repeaters?
  A. AM
  B. SSB
  C. PSK
  D. FM

Since VHF and Ultra High Frequency (UHF) links tend to be line-of-sight or short-distance hops, the background noise is not the issue, as in the previous question. Now we can go for higher quality. While all the choices are possible, current amateur practice is that FM is the mode of choice, so **Answer D** is the right choice. The others are not common in current practice in this mode.

T8A05 Which of the following types of emission has the narrowest bandwidth?
  A. FM voice
  B. SSB voice
  C. CW
  D. Slow-scan TV

If we arrange the choices in order of increasing bandwidth, we would obtain: Continuous Wave (CW) (150 Hz), SSB (3 kHz), Slow-scan TV (3 kHz), and FM (15 kHz). Therefore, the correct answer for this question is **Answer C**, since CW has the narrowest bandwidth.

T8A06 Which sideband is normally used for 10 meter HF, VHF and UHF single-sideband communications?
  A. Upper sideband
  B. Lower sideband
  C. Suppressed sideband
  D. Inverted sideband

This is a case where you will need to learn the amateur usage convention. Your rig will allow you to use either the USB or LSB modes. By convention, transmissions above 10 meters use USB, so **Answer A** is correct. Answers C and D are technobabble terms to distract you.

T8A07 What is an advantage of single sideband (SSB) over FM for voice transmissions?
  A. SSB signals are easier to tune
  B. SSB signals are less susceptible to interference
  C. SSB signals have narrower bandwidth
  D. All of these choices are correct

Based on the previous questions, you may be able to spot the right answer here. The correct choice is the narrower bandwidth, as in **Answer C**. Answers A and B are technically incorrect statements, so they are not good choices. This also makes Answer D incorrect.

T8A08 What is the approximate bandwidth of a single sideband (SSB) voice signal?
A. 1 kHz
B. 3 kHz
C. 6 kHz
D. 15 kHz

This is one of those facts that you will need to learn: SSB voice typically occupies around 3 kHz, so **Answer B** is the right choice. Be careful with Answer C because it contains 6 kHz, which is the bandwidth needed for DSB voice. Answer D has the bandwidth for FM voice transmissions. Answer A is to distract you.

T8A09 What is the approximate bandwidth of a VHF repeater FM phone signal?
A. Less than 500 Hz
B. About 150 kHz
C. Between 10 and 15 kHz
D. Between 50 and 125 kHz

You should be able to spot the 10 to 15 kHz in **Answer C** as the right choice, based on the previous questions.

T8A10 What is the typical bandwidth of analog fast-scan TV transmissions on the 70 cm band?
A. More than 10 MHz
B. About 6 MHz
C. About 3 MHz
D. About 1 MHz

Fast-scan TV uses the same bandwidth as regular analog TV (not HDTV). **Answer B** has the correct answer: about 6 MHz.

T8A11 What is the approximate maximum bandwidth required to transmit a CW signal?
A. 2.4 kHz
B. 150 Hz
C. 1000 Hz
D. 15 kHz

Based on the previous questions, you should be able to spot **Answer B** as the right choice: around 150 Hz. Answer A is for SSB phone, and Answer D is for FM phone.

8.4 T8B – Amateur Satellite Operation

8.4.1 Overview

The *Amateur Satellite Operation* question group in Subelement T8 introduces you to amateur space station operating concepts. The *Amateur Satellite Operation* group covers topics such as

- Amateur satellite operation
- Doppler shift
- Basic orbits
- Operating protocols
- Transmitter power considerations
- Telemetry and telecommand
- Satellite tracking

There is a total of 12 questions in this group of which one will be selected for the exam.

8.4.2 Questions

T8B01 What telemetry information is typically transmitted by satellite beacons?

A. The signal strength of received signals
B. Time of day accurate to plus or minus 1/10 second
C. Health and status of the satellite
D. All of these choices are correct

Telemetry data are measurements from the satellite, usually about the current state of the electronics within the satellite. Engineers call this health and status information making **Answer C** the correct choice. A Global Positioning System (GPS) satellite gives very exact timing information, so there is no need for a satellite beacon for this information. The satellite's designer may embed a signal strength measurement in the status telemetry data, so Answer A is not the best choice. Since Answers A and B are incorrect, Answer D is also incorrect.

T8B02 What is the impact of using too much effective radiated power on a satellite uplink?

A. Possibility of commanding the satellite to an improper mode
B. Blocking access by other users
C. Overloading the satellite batteries
D. Possibility of rebooting the satellite control computer

Satellite transceiver radio inputs are sensitive to signal levels. If a signal is too strong, it can block out other users and deny them service. This makes **Answer B** the correct choice. It will not affect batteries, so Answer C is incorrect. If the signal places the satellite in an improper mode or causes a computer reboot, then the satellite's design

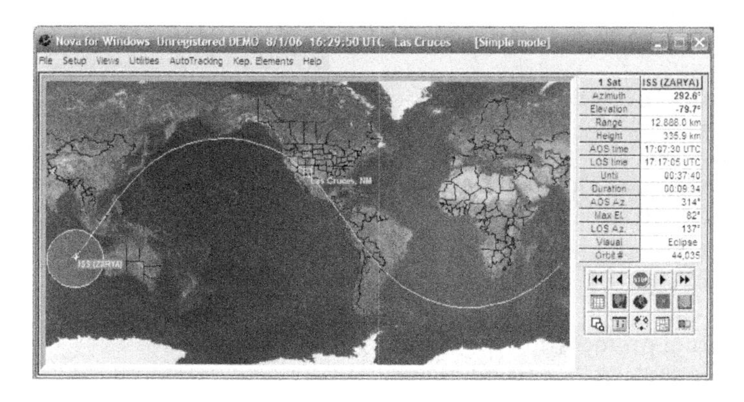

Figure 8.4: Satellite tracking program user display for real-time satellite tracking.

is improper, and these are here to distract you.

T8B03 Which of the following are provided by satellite tracking programs?

A. Maps showing the real-time position of the satellite track over the earth
B. The time, azimuth, and elevation of the start, maximum altitude, and end of a pass
C. The apparent frequency of the satellite transmission, including effects of Doppler shift
D. All of these answers are correct

A good satellite tracking program will give the user information about each of the items in Answers A, B, and C, which makes **Answer D** the best choice. Figure 8.4 shows an example of a satellite tracking display.

T8B04 What mode of transmission is commonly used by amateur radio satellites?

A. SSB
B. FM
C. CW/data
D. All of these choices are correct

If you become active with the amateur satellite community, you will find operators using each of the modes of Answers A, B, and C. This makes **Answer D** the best choice for this question.

T8B05 What is a satellite beacon?

A. The primary transmit antenna on the satellite
B. An indicator light that that shows where to point your antenna
C. A reflective surface on the satellite
D. A transmission from a satellite that contains status information

Modern satellites use antennas, and not reflective surfaces to relay signals, so Answer C is ancient history. A beacon is for information and is not an antenna system, so Answer A is incorrect. Answer B might be helpful, but it would need to be a really bright light to be visible from space, so this is a distraction. The correct choice is the informational transmission given in **Answer D**.

T8B06 Which of the following are inputs to a satellite tracking program?

A. The weight of the satellite
B. The Keplerian elements
C. The last observed time of zero Doppler shift
D. All of these answers are correct

While the mass of a satellite and times of zero Doppler shift have their place, they are not normal input parameters to satellite tracking programs, so Answers A and C are incorrect choices. This makes Answer D incorrect as well. Tracking programs do, however, need the Keplerian elements, as in **Answer B**, making that the correct choice. The Two Line Elements (TLE) for the satellite encodes the Keplerian elements for the program.

T8B07 With regard to satellite communications, what is Doppler shift?

A. A change in the satellite orbit
B. A mode where the satellite receives signals on one band and transmits on another
C. An observed change in signal frequency caused by relative motion between the satellite and the earth station
D. A special digital communications mode for some satellites

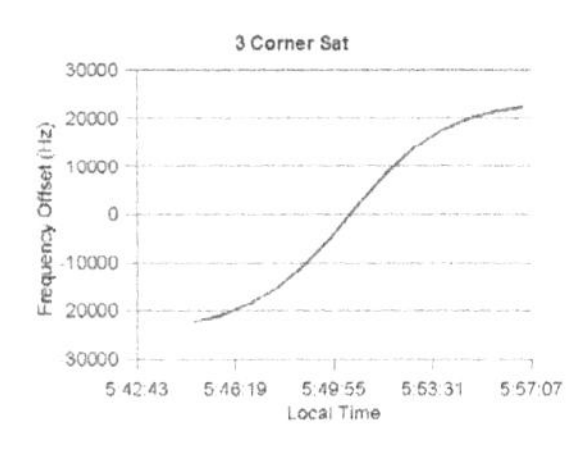

Figure 8.5: The observed carrier Doppler shift for an orbiting satellite during a contact.

The Doppler effect is a frequency change in the radio signal due to motion, as shown in Figure 8.5, so **Answer C** is the right choice. Answer D is technobabble. Answer B is a description of "pacsat" mode on transceivers, so it is incorrect here. Answer A may cause a Doppler shift under certain circumstances, but the two are not the same, so this is also incorrect.

Table 8.2: Amateur Satellite Communication Modes

Mode	Satellite Receiving Band	Authorized Frequencies (MHz)	Satellite Transmitting Band	Authorized Frequencies (MHz)
V/H	VHF	144.30 – 144.50 145.80 – 146.00	HF	29.300 – 29.510
U/V	UHF	435.00 – 438.00	VHF	144.30 – 144.50 145.80 – 146.00
V/U	VHF	144.30 – 144.50 145.80 – 146.00	UHF	435.00 – 438.00
L/U	L-Band	1260 – 1270	UHF	435.00 – 438.00

T8B08 What is meant by the statement that a satellite is operating in mode U/V?

A. The satellite uplink is in the 15 meter band and the downlink is in the 10 meter band
B. The satellite uplink is in the 70 cm band and the downlink is in the 2 meter band
C. The satellite operates using ultraviolet frequencies
D. The satellite frequencies are usually variable

In satellite operations, there are several possible uplink and downlink communications operating modes for the satellites, as Table 8.2 shows. The first letter of the mode designator is the receiving band, and the second letter is the transmitting band. These designations are from the point of view of the satellite, and not the earth station using the satellite. Answers C and D are silly distractors. **Answer B** captures the sense of the communications modes, so be sure to choose this answer. Answer A is the opposite sense.

T8B09 What causes spin fading when referring to satellite signals?

A. Circular polarized noise interference radiated from the sun
B. Rotation of the satellite and its antennas
C. Doppler shift of the received signal
D. Interfering signals within the satellite uplink band

Many satellites spin to maintain a stable orientation. However, this spin can cause problems in antenna pointing, and engineers call the resulting change in signal strength "spin fading." **Answer B** captures this definition. Answer A is technobabble. The Doppler shift in Answer C is due to the satellite's orbital motion and not rotation, so it is incorrect. Answer D is technically incorrect.

T8B10 What do the initials LEO tell you about an amateur satellite?

A. The satellite battery is in Low Energy Operation mode
B. The satellite is performing a Lunar Ejection Orbit maneuver
C. The satellite is in a Low Earth Orbit
D. The satellite uses Light Emitting Optics

In the satellite community, this is Low Earth Orbit (LEO), as in **Answer C**. The other choices are silly distractors.

T8B11 Who may receive telemetry from a space station?

A. Anyone who can receive the telemetry signal
B. A licensed radio amateur with a transmitter equipped for interrogating the satellite
C. A licensed radio amateur who has been certified by the protocol developer
D. A licensed radio amateur who has registered for an access code from AMSAT

Each of the options given in Answers B, C, and D may sound reasonable, but they are incorrect statements. The general rule of good amateur practice is that anyone may receive a transmission. This makes **Answer A** the correct choice.

T8B12 Which of the following is a good way to judge whether your uplink power is neither too low nor too high?

A. Check your signal strength report in the telemetry data
B. Listen for distortion on your downlink signal
C. Your signal strength on the downlink should be about the same as the beacon
D. All of these choices are correct

Your signal strength may not appear in telemetry nor may the satellite have a beacon, so Answers A and C are not going to help. Since these are incorrect, Answer D is also incorrect. Listening to your downlink signal will be a good indicator of signal strength making **Answer B** the correct choice.

8.5 T8C – Operating Activities

8.5.1 Overview

The *Operating Activities* question group in Subelement T8 introduces you to amateur radio operating activities beyond phone and CW messages. The *Operating Activities* group covers topics such as

- Radio direction finding
- Radio control
- Contests
- Linking over the Internet
- Grid locators

There is a total of 11 questions in this group of which one will be selected for the exam.

8.5.2 Questions

T8C01 Which of the following methods is used to locate sources of noise interference or jamming?

A. Echolocation
B. Doppler radar
C. Radio direction finding
D. Phase locking

One way to locate interference sources is to use the same techniques as used in a "fox hunt" or a hidden transmitter hunt. To do this, you will need a set of radio direction finding equipment. This technique matches **Answer C**. Echolocation is fine if you are looking for submarines or are a bat, but not for radios. Scientists use a Doppler radar for weather tracking, not radios. Phase locking is a receiver technique that does not assist in pinpointing a location.

T8C02 Which of these items would be useful for a hidden transmitter hunt?

A. Calibrated SWR meter
B. A directional antenna
C. A calibrated noise bridge
D. All of these choices are correct

A radio direction finding set-up requires a directional antenna to point the way to the transmitter, so **Answer B** is the best choice among those given. A Standing Wave Ratio (SWR) meter and a noise bridge are good devices to have in your shack to maximize the performance of your equipment, but they will not point the direction to a hidden antenna, so they are not good choices here. Since Answers A and C are incorrect, Answer D must be incorrect as well.

T8C03 What popular operating activity involves contacting as many stations as possible during a specified period of time?

A. Contesting
B. Net operations
C. Public service events
D. Simulated emergency exercises

This question is giving an example of contesting, so **Answer A** is correct. A popular contest is Field Day each June. However, there are many types of contests for different operating modes. Net operations are usually regular radio meetings among group members for social or information sharing, so this is not a good choice. Operators do not use Answers C and D for the specific purpose of contacting as many stations as possible, so they are incorrect choices.

T8C04 Which of the following is good procedure when contacting another station in a radio contest?

A. Sign only the last two letters of your call if there are many other stations calling
B. Contact the station twice to be sure that you are in his log
C. Send only the minimum information needed for proper identification and the contest exchange
D. All of these choices are correct

Many contests require that the operators exchange specific information to verify that they have made a proper contest contact. Since the contest is a fixed duration and the time for an exchange is usually short, the operators should only exchange the required information, as in **Answer C**. Answer B is a waste of contest time, so this is a bad choice. Answer A does not guarantee exchanging contest information, so this is not a good choice. Since Answers A and B are incorrect, Answer D is also incorrect.

T8C05 What is a grid locator?

A. A letter-number designator assigned to a geographic location
B. A letter-number designator assigned to an azimuth and elevation
C. An instrument for neutralizing a final amplifier
D. An instrument for radio direction finding

The amateur community has a world-wide grid to assist operators in specifying their location in a more precise manner. The grid locator is two letters followed by two numbers followed by an optional two or more letters and numbers. For example, "FM17td" is the author's location. This makes **Answer A** the correct choice for this question. You can find a Web-based calculator at `http://www.levinecentral.com/ham/grid_square.php`. Answer B seems logical, but it is technically incorrect. Many power amplifier tubes have grids, but that is not the sense of this question. Aiding in radio direction finding may sound reasonable after the earlier questions, but it is not the sense of this question.

T8C06 How is access to some IRLP nodes accomplished?

A. By obtaining a password that is sent via voice to the node
B. By using DTMF signals
C. By entering the proper internet password
D. By using CTCSS tone codes

Unless you have used the Internet Radio Linking Project (IRLP) protocol, this question may not make much sense, so you will just need to memorize the right answer. The correct method is to enter the IRLP node numbers via a keypad's Dual Tone Multifrequency (DTMF) tones, so **Answer B** is the right choice. You can find out more about IRLP by Web searching at `http://www.irlp.net/`.

T8C07 What is meant by Voice Over Internet Protocol (VoIP) as used in amateur radio?

A. A set of rules specifying how to identify your station when linked over the internet to another station
B. A set of guidelines for contacting DX stations during contests using internet access
C. A technique for measuring the modulation quality of a transmitter using remote sites monitored via the internet
D. A method of delivering voice communications over the internet using digital techniques

Voice over IP (VoIP) is for voice communications over the Internet using digital methods, as in Answer D. The other choices are to distract you.

T8C08 What is the Internet Radio Linking Project (IRLP)?

A. A technique to connect amateur radio systems, such as repeaters, via the internet using Voice Over Internet Protocol (VoIP)
B. A system for providing access to websites via amateur radio
C. A system for informing amateurs in real time of the frequency of active DX stations
D. A technique for measuring signal strength of an amateur transmitter via the internet

The IRLP is an activity enabled by VoIP making **Answer A** the correct choice. The other options might be nice applications to have, but they are distractions for you.

T8C09 How might you obtain a list of active nodes that use VoIP?

A. By subscribing to an on line service
B. From on line repeater lists maintained by the local repeater frequency coordinator
C. From a repeater directory
D. All of these choices are correct

All the options given in Answers A, B, and C are places for you to go to for lists of active nodes using VoIP. This makes **Answer D** the best choice for this question.

T8C10 What must be done before you may use the EchoLink system to communicate using a repeater?

A. You must complete the required EchoLink training
B. You must have purchased a license to use the EchoLink software
C. You must be sponsored by a current EchoLink user
D. You must register your call sign and provide proof of license

Each of the options in Answers A, B, and C may sound reasonable, but they are distractions. Registration and proof of license are all that EchoLink requires, so

Answer D is the right choice.

T8C11 What name is given to an amateur radio station that is used to connect other amateur stations to the internet?

A. A gateway
B. A repeater
C. A digipeater
D. A beacon

To attach your station to the Internet, you will need a gateway, as in **Answer A**. A repeater and a digipeater both retransmit radio signals, so these are incorrect. A beacon station is for measuring radio propagation conditions, so it is also an incorrect choice.

8.6 T8D – Non-voice and Digital Communications

8.6.1 Overview

The *Non-voice and Digital Communications* question group in Subelement T8 introduces you to additional non-voice amateur radio activities. The *Non-voice and Digital Communications* group covers topics such as

- Image signals
- Digital modes
- CW
- Packet radio
- PSK31
- APRS
- Error detection and correction
- NTSC
- Amateur radio networking
- Digital Mobile/Migration Radio

There is a total of 14 questions in this group of which one will be selected for the exam.

8.6.2 Questions

T8D01 Which of the following is a digital communications mode?

A. Packet radio
B. IEEE 802.11
C. JT65
D. All of these choices are correct

Each of the modes listed in Answers A, B, and C are examples of digital communications, so **Answer D** is the best choice to answer this question.

T8D02 What does the term "APRS" mean?
A. Automatic Packet Reporting System
B. Associated Public Radio Station
C. Auto Planning Radio Set-up
D. Advanced Polar Radio System

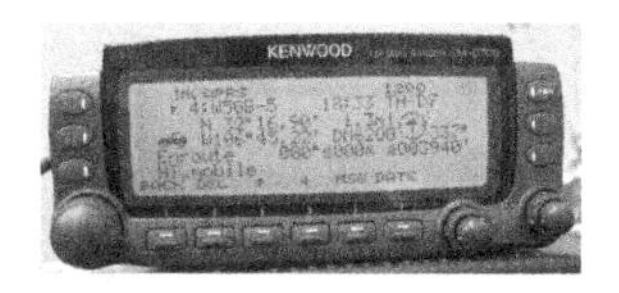

Figure 8.6: An example APRS message on a receiver display.

This is the Automatic Packet Reporting System (APRS), as in **Answer A** (generally pronounced as "APE-ers"). The other choices are to distract you.

T8D03 Which of the following devices is used to provide data to the transmitter when sending automatic position reports from a mobile amateur radio station?
A. The vehicle speedometer
B. A WWV receiver
C. A connection to a broadcast FM sub-carrier receiver
D. A Global Positioning System receiver

The APRS system works with a GPS receiver to generate the position locations (although the user can also enter the information manually). This makes **Answer D** the right choice. A WWV receiver is good for getting time signals, but not position. The other choices are distractions.

T8D04 What type of transmission is indicated by the term "NTSC"?
A. A Normal Transmission mode in Static Circuit
B. A special mode for earth satellite uplink
C. An analog fast scan color TV signal
D. A frame compression scheme for TV signals

The National Television System Committee (NTSC) refers to part of the official name for the standard analog color television signal format (also known as "fast scan"). This makes **Answer C** the correct choice. The others are silly distractors.

T8D05 Which of the following is an application of APRS (Automatic Packet Reporting System)?

A. Providing real time tactical digital communications in conjunction with a map showing the locations of stations
B. Showing automatically the number of packets transmitted via PACTOR during a specific time interval
C. Providing voice over internet connection between repeaters
D. Providing information on the number of stations signed into a repeater

A common display for APRS shows the locations of participating stations along with the associated message traffic, which makes **Answer A** the right choice. APRS is not involved with PACTOR or VoIP, so Answers B and C are not good choices. It will also not help with Answer D.

T8D06 What does the abbreviation "PSK" mean?

A. Pulse Shift Keying
B. Phase Shift Keying
C. Packet Short Keying
D. Phased Slide Keying

By now you should be able to spot that **Answer B** correctly describes the term PSK. The other definitions are just to see if you have learned this new term.

T8D07 Which of the following best describes DMR (Digital Mobile Radio)?

A. A technique for time-multiplexing two digital voice signals on a single 12.5 kHz repeater channel
B. An automatic position tracking mode for FM mobiles communicating through repeaters
C. An automatic computer logging technique for hands-off logging when communicating while operating a vehicle
D. A digital technique for transmitting on two repeater inputs simultaneously for automatic error correction

Digital Mobile Radio (DMR) is a protocol that allows two voice channels to share a repeater channel, as in **Answer A**. The other choices are to distract you.

T8D08 Which of the following may be included in packet transmissions?

A. A check sum that permits error detection
B. A header that contains the call sign of the station to which the information is being sent
C. Automatic repeat request in case of error
D. All of these choices are correct

A packet transmission may include each of the items listed in Answers A, B, and C, so **Answer D** is the best choice to answer this question.

T8D09 What code is used when sending CW in the amateur bands?

A. Baudot
B. Hamming
C. International Morse
D. All of these choices are correct

CW is a quick way to say "International Morse Code," so **Answer C** is correct. The other answers are digital codes, which operators do not use with CW.

T8D10 Which of the following operating activities is supported by digital mode software in the WSJT suite?

A. Moonbounce or Earth-Moon-Earth
B. Weak-signal propagation beacons
C. Meteor scatter
D. All of these choices are correct

The WSJT software suite specializes in weak-signal modes. Each sending technique listed in Answers A, B, and C is a weak-signal mode, so **Answer D** is the best choice to answer this question.

T8D11 What is an ARQ transmission system?

A. A special transmission format limited to video signals
B. A system used to encrypt command signals to an amateur radio satellite
C. A digital scheme whereby the receiving station detects errors and sends a request to the sending station to retransmit the information
D. A method of compressing the data in a message, so more information can be sent in a shorter time

The Automatic Repeat reQuest (ARQ) protocol is a packet protocol to automatically request retransmissions of packets received with transmission errors, so **Answer C** is correct. The protocol does not compress or encrypt the data, so Answers B and D are technically incorrect. Transmitters that stream video signals do not use ARQ, so Answer A is also incorrect.

T8D12 (A) Which of the following best describes Broadband-Hamnet(TM), also referred to as a high-speed multi-media network?

A. An amateur-radio-based data network using commercial Wi-Fi gear with modified firmware
B. A wide-bandwidth digital voice mode employing DRM protocols
C. A satellite communications network using modified commercial satellite TV hardware
D. An internet linking protocol used to network repeaters

Broadband Hamnet is a recent mode that has become popular and it is a data network using modified Wi-Fi equipment as in **Answer A**. The other options are to distract you.

T8D13 What is FT8?

A. A wideband FM voice mode
B. A digital mode capable of operating in low signal-to-noise conditions that transmits on 15-second intervals
C. An eight channel multiplex mode for FM repeaters
D. A digital slow scan TV mode with forward error correction and automatic color compensation

FT8 is a relatively new digital mode especially designed for weak signal conditions making **Answer B** the correct choice. The other choices are to distract you.

T8D14 What is an electronic keyer?

A. A device for switching antennas from transmit to receive
B. A device for voice activated switching from receive to transmit
C. A device that assists in manual sending of Morse code
D. An interlock to prevent unauthorized use of a radio

An electronic keyer helps the operator in sending CW as in **Answer C**. Answer A is a "T/R" switch. Answer B is a Voice Operated Switch (VOX) switch. Answer D is another distraction choice.

Chapter 9

T9 – ANTENNAS AND FEED LINES

9.1 Introduction

This chapter goes into more details about the antennas and feed lines that have been coming up in earlier chapters. As you may have surmised from the earlier chapters, the antenna is responsible for emitting the Radio Frequency (RF) energy from the transmitter and picking up the RF energy for the receiver. Usually, the antenna works equally well in the transmission mode as in the reception mode. This *Antennas and Feed Lines* subelement has the following question groups:

A. Antennas
B. Feed lines

This will generate two questions on the Technician examination.

9.2 Radio Engineering Concepts

Antenna Basics The antenna questions at the Technician level are more descriptive than quantitative, so there will only be a few equations to master. First, we need to learn a few fundamental antenna descriptions

Isotropic — a theoretically ideal antenna that radiates the RF energy equally in all directions

Omnidirectional — an antenna that radiates a significant amount of RF energy in all directions, but not as uniformly as an isotropic

Hemispherical — an antenna that radiates nearly all its energy in a hemisphere

Beam or Directional — an antenna that radiates nearly all its energy in a relatively confined direction

Figure 9.1 illustrates several real antennas. Not shown in the figure is a *dipole* antenna, which is a wire conductor, usually mounted horizontally, designed to

(a) A vertical antenna.

(b) Two Yagi antennas for different frequency bands.

(c) A 1-m Ka-Band parabolic dish antenna with an offset feed.

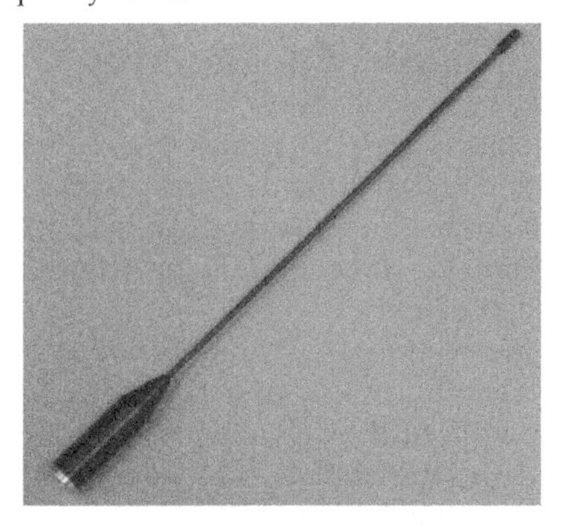

(d) A "rubber duck" antenna for UHF frequencies.

Figure 9.1: Pictures of representative antenna types.

operate at a given frequency. Other types you will encounter include

Vertical — a vertically-oriented conductor, usually shaped like a tube, which designers often configure to operate on multiple bands

Rubber Duck — a flexible antenna, which designers intend for use usually on hand-held radios

Yagi — a beam antenna with a single driven element and multiple "parasitic" elements designed to give relatively high gain, and designed for a single band

Dish — a parabolic dish with high gain, and designed for a single band

There are other antenna designs that you will encounter as you progress through the licensing process.

The two most common antenna characterizations are the

Gain — the strength of the antenna pattern relative to a reference antenna, such as an isotropic antenna; usually the manufacturer lists only the maximum gain

Beamwidth — the angular spread of the antenna pattern at the maximum gain; usually, the manufacturers use Half Power Beam Width (HPBW), which measures the angular extent of the pattern at one-half of the maximum gain

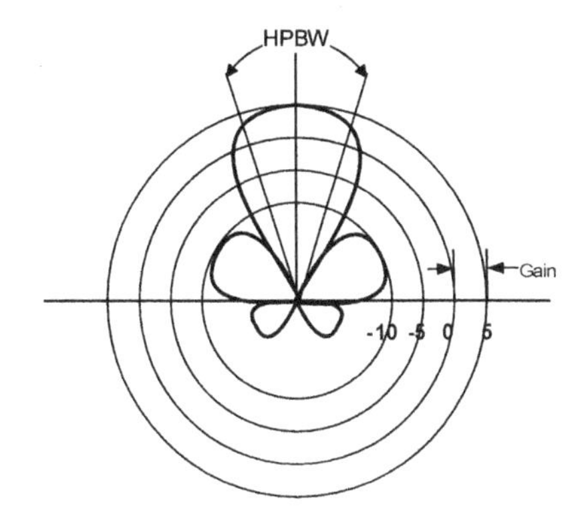

Figure 9.2: Representative antenna pattern with gain and bandwidth measures

Figure 9.2 shows these relative to the antenna pattern measurement. Note: the antenna's gain and beamwidth are related quantities. If the beamwidth is large, the gain is small, while if the beamwidth is small, the gain is large.

Manufacturers design conductor antennas, such as dipoles, based on a multiple of the operating frequency. Typical design equations for the length of these antennas in feet, L, based on the operating frequency, f, in MHz are

One Wavelength Antenna — $L = 936/f$

Half Wavelength Antenna — $L = 936/(2 * f) = 468/f$

Quarter Wavelength Antenna — $L = 936/(4 * f) = 234/f$

Notice that you only need to know the first equation, and then divide the result by either 2 or 4, as appropriate.

Feed Lines As we saw in Chapter 7, the Standing Wave Ratio (SWR) on the feed line is important for correct station operation. A low SWR, preferably 1:1, implies a good match between the components. A good match means that the maximum RF power is transferred and does not go into heating the feed line. Most current amateur radio transceivers and antennas have an impedance of 50 Ω. This implies that the feed line should also have an impedance of 50 Ω to maximize power transfer.

Suppose your antenna does not have an impedance of 50 Ω, is there something that you can do to correct this problem? One way to help the situation is with an *antenna tuner*. Many transceivers have built-in antenna tuners to give a better match if the antenna is not exactly 50 Ω at the operating frequency. Usually, these built-in tuners have a limited range of tuning. If you need more, an external antenna tuner, like the one shown in Figure 4.2 is very helpful in enabling a wider range of operations.

Co-axial cables do not behave identically at all frequencies. In particular, the amount of cable loss depends on the frequency. Generally, the higher the frequency, the more loss per foot of length. The amateur radio operator needs to take this loss in consideration when planning the location for the antenna. Related to this, is the connector on the cable. These connectors also have losses depending upon the operating frequency. Most transceivers use a Ultra High Frequency (UHF) (or a PL-259 type) connector. Hand-held radios tend to use BNC or SMA connectors. You need to match the co-axial connector type to the radio and antenna connector types.

9.3 T9A – Antennas

9.3.1 Overview

The *Antennas* question group in Subelement T9 introduces you to more details on antenna concepts used in amateur radio. The *Antennas* group covers topics such as

- Vertical and horizontal polarization
- Concept of gain
- Common portable and mobile antennas
- Relationships between antenna length and frequency
- Concept of dipole antennas

There is a total of 12 questions in this group of which one will be selected for the exam.

9.3.2 Questions

T9A01 What is a beam antenna?

A. An antenna built from aluminum I-beams
B. An omnidirectional antenna invented by Clarence Beam
C. An antenna that concentrates signals in one direction
D. An antenna that reverses the phase of received signals

Electrically, you can use an I-beam for an antenna. However, an antenna made of I-beams would be really big, and the neighbors might raise a ruckus, so this answer is a silly choice. An antenna that transmits in all directions is known as an omnidirectional antenna, and not a beam antenna. Clarence Beam is known as a federal judge, and not an antenna designer. An antenna that concentrates the transmission in one direction is a beam antenna, so **Answer C** is the right choice. The beam antenna does not reverse the phase of the transmission, so Answer D is also incorrect.

T9A02 Which of the following describes a type of antenna loading?

A. Inserting an inductor in the radiating portion of the antenna to make it electrically longer
B. Inserting a resistor in the radiating portion of the antenna to make it resonant
C. Installing a spring in the base of a mobile vertical antenna to make it more flexible
D. Strengthening the radiating elements of a beam antenna to better resist wind damage

Antenna loading is a technique that antenna designers use to make the antenna act as close to a full-size antenna as possible by inserting an inductor to make it seem electrically longer. **Answer A** describes this technique, so this is the correct choice. The resistor of Answer B is not reactive, so this will not help electrically. The options of Answers C and D have to do with mechanical construction and not matching, so they are distraction answers.

T9A03 Which of the following describes a simple dipole oriented parallel to the Earth's surface?

A. A ground-wave antenna
B. A horizontally polarized antenna
C. A rhombic antenna
D. A vertically polarized antenna

If the conductor is parallel to the Earth, then it is horizontal, and the electric field also has a horizontal orientation. This makes **Answer B** the correct choice and all others technically incorrect.

T9A04 What is a disadvantage of the "rubber duck" antenna supplied with most handheld radio transceivers when compared to a full-sized quarter-wave antenna?

A. It does not transmit or receive as effectively
B. It transmits a circularly polarized signal
C. If the rubber end cap is lost, it will unravel very easily
D. All of these choices are correct

Answer B is untrue because duckies do not transmit circularly polarized signals. Answer C is also untrue; they will become unsafe, but not necessarily unwind. Since Answers B and C are untrue, Answer D is also incorrect. The disadvantage to a ducky is that they are not effective transmitters and receivers of the radio signal, so **Answer A** is the right choice.

T9A05 How would you change a dipole antenna to make it resonant on a higher frequency?

A. Lengthen it
B. Insert coils in series with radiating wires
C. Shorten it
D. Add capacitive loading to the ends of the radiating wires

The resonant frequency of a dipole antenna is related to the length of the antenna. As the length increases, the resonant frequency goes down. Therefore, to make the resonant frequency higher, one must make the dipole antenna shorter. **Answer C** is the best choice.

T9A06 What type of antennas are the quad, Yagi, and dish?

A. Non-resonant antennas
B. Log periodic antennas
C. Directional antennas
D. Isotropic antennas

Each of the options in Answers A, B, and D are untrue statements. However, these are directional or beam antennas, so **Answer C** is the right choice.

T9A07 What is a disadvantage of using a handheld VHF transceiver, with its integral antenna, inside a vehicle?

A. Signals might not propagate well due to the shielding effect of the vehicle
B. It might cause the transceiver to overheat
C. The SWR might decrease, decreasing the signal strength
D. All of these choices are correct

The metal in the car's body car can block a good deal of the radio signal from exiting the vehicle. This will make the signal weaker inside than outside of the vehicle, as in **Answer A**. If your transceiver over heats, you have other problems, so Answer B is not a good choice. Answer C is technically not correct. Since Answers B and C are incorrect, Answer D is also incorrect.

T9A08 What is the approximate length, in inches, of a quarter-wavelength vertical antenna for 146 MHz?

A. 112
B. 50
C. 19
D. 12

Here we use the name of the antenna to compute its length. A good design equation for ¼-wavelength antennas is $L(\text{ft}) = 234/f(\text{MHz})$. Using the given frequency of 146 MHz, we get $L(\text{ft}) = 234/146 = 1.6$ ft = 19 in. This makes **Answer C** correct.

T9A09 What is the approximate length, in inches, of a half-wavelength 6 meter dipole antenna?

A. 6
B. 50
C. 112
D. 236

This is like the previous question. The design equation for the ½-wavelength antenna is now $L(\text{ft}) = 468/f(\text{MHz})$ (twice the length of the antenna in the previous question). 6 m is approximately 50 MHz, and we can estimate the length as $L(\text{ft}) = 468/50 = 9.36$ ft or 112 inches. **Answer C** is the right choice.

T9A10 In which direction does a half-wave dipole antenna radiate the strongest signal?

A. Equally in all directions
B. Off the ends of the antenna
C. Broadside to the antenna
D. In the direction of the feed line

This is probably an answer you will need to memorize. For a half-wave dipole, the radiation comes radially out from it, or broadside to the antenna (think of the antenna as a pencil placed through the hole in a donut with the radiation being the donut). **Answer C** is the right description.

T9A11 What is the gain of an antenna?

A. The additional power that is added to the transmitter power
B. The additional power that is lost in the antenna when transmitting on a higher frequency
C. The increase in signal strength in a specified direction compared to a reference antenna
D. The increase in impedance on receive or transmit compared to a reference antenna

The gain is a measure of the relative signal strength in a specified direction relative to a reference antenna, as defined in **Answer C**. While the gain deals with antenna power, it is a measurement relative to the reference antenna, so Answers A and B are incorrect. It is not an impedance measurement, so Answer D is also incorrect.

T9A12 What is an advantage of using a properly mounted 5/8 wavelength antenna for VHF or UHF mobile service?

A. It has a lower radiation angle and more gain than a 1/4 wavelength antenna
B. It has very high angle radiation for better communicating through a repeater
C. It eliminates distortion caused by reflected signals
D. It has 10 times the power gain of a 1/4 wavelength design

Generally, when operating mobile, we wish the RF signal to go out in as a horizontal direction as possible, for example to hit local repeaters. This is what "low angle" radiation implies. **Answer A** matches this desired behavior with some extra gain, so this is the right choice. The other choices are technically incorrect statements.

9.4 T9B – Feed Lines

9.4.1 Overview

The *Feed Lines* question group in Subelement T9 introduces you to more information on antenna feed lines. The *Feed Lines* group covers topics such as

- Types
- Attenuation vs. frequency
- Selecting
- SWR concepts
- Antenna tuners (couplers)
- RF connectors: selecting and weather protection

There is a total of 11 questions in this group of which one will be selected for the exam.

9.4.2 Questions

T9B01 Why is it important to have low SWR when using coaxial cable feed line?
A. To reduce television interference
B. To reduce signal loss
C. To prolong antenna life
D. All of these choices are correct

As we indicated earlier, a low SWR implies a good impedance match. A good impedance match implies the most efficient transfer of power and least line loss, so **Answer B** is the correct reasoning for this question. A low SWR will not affect TV interference nor will it affect antenna lifetime. Since Answers A and C are incorrect, Answer D is also incorrect.

T9B02 What is the impedance of most coaxial cables used in amateur radio installations?
A. 8 ohms
B. 50 ohms
C. 600 ohms
D. 12 ohms

For best power transfer conditions, the impedances should be as close to equal (matched) as possible. The output of the transceiver and the input of the antenna commonly have impedances of 50 Ω. This means that the co-ax should also be 50 Ω, so **Answer B** is the correct response. The other choices give mismatches.

T9B03 Why is coaxial cable the most common feed line selected for amateur radio antenna systems?

A. It is easy to use and requires few special installation considerations
B. It has less loss than any other type of feed line
C. It can handle more power than any other type of feed line
D. It is less expensive than any other types of feed line

The statements given in Answers B, C, and D are all untrue. The ease of use and installation of co-axial cable are true statements, so **Answer A** is the right choice.

T9B04 What is the major function of an antenna tuner (antenna coupler)?

A. It matches the antenna system impedance to the transceiver's output impedance
B. It helps a receiver automatically tune in weak stations
C. It allows an antenna to be used on both transmit and receive
D. It automatically selects the proper antenna for the frequency band being used

Answer B sounds like a useful device, but that is a function usually performed by the station operator (you). Impedance matching is the true function of an antenna tuner, as found in **Answer A**, so that is the right choice here. Operators usually call Answer C a T-R switch, but one typically does not have both a transceiver and a receiver in the same setup. Answer D describes an antenna switch.

T9B05 In general, what happens as the frequency of a signal passing through coaxial cable is increased?

A. The characteristic impedance decreases
B. The loss decreases
C. The characteristic impedance increases
D. The loss increases

The loss in a co-axial cable increases with frequency, so **Answer D** is the right choice. The other choices are not intrinsic electrical properties of co-axial cables.

T9B06 Which of the following connectors is most suitable for frequencies above 400 MHz?

A. A UHF (PL-259/SO-239) connector
B. A Type N connector
C. An RS-213 connector
D. A DB-25 connector

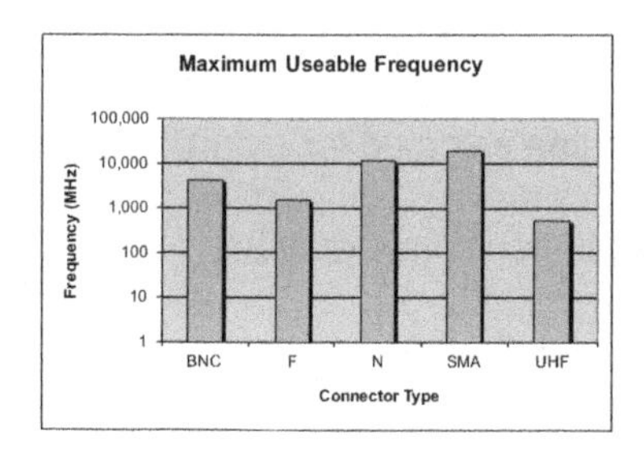

Figure 9.3: Usable frequency ranges for typical RF connectors.

As we can see from Figure 9.3, the N-type and SMA connectors have the best performance (highest maximum usable frequency), so we chose **Answer B** as the best choice among those given. The RS-213 is a fictitious connector, while the DB-25 is a serial data connector, so neither are good choices.

T9B07 Which of the following is true of PL-259 type coax connectors?

A. They are preferred for microwave operation
B. They are water tight
C. They are commonly used at HF frequencies
D. They are a bayonet type connector

Users also call PL-259 connectors UHF connectors, but do not let the name fool you. As Figure 9.3 shows, they are better for High Frequency (HF) and Very High Frequency (VHF) than they are for UHF. **Answer C** is the best choice among those given here. Answers A and B are incorrect statements.

T9B08 Why should coax connectors exposed to the weather be sealed against water intrusion?

A. To prevent an increase in feed line loss
B. To prevent interference to telephones
C. To keep the jacket from becoming loose
D. All of these choices are correct

Water in the feed line will increase the loss, so one tries to keep water out of the connectors. **Answer A** is the correct reason. Answers B and C will not happen by weather sealing the connectors, so they are incorrect choices. That also makes Answer D incorrect.

T9B09 What might cause erratic changes in SWR readings?

A. The transmitter is being modulated
B. A loose connection in an antenna or a feed line
C. The transmitter is being over-modulated
D. Interference from other stations is distorting your signal

Modulating the transmitter will not change the SWR in a properly designed rig, so Answer A is not a good choice. Over-modulation will distort the signal, but not change the SWR either, so Answer C is incorrect. Answer D is just a distraction. However, loose connections will change the apparent impedance of a device, and change the SWR, so **Answer B** is the right choice to answer this question.

T9B10 What is the electrical difference between RG-58 and RG-8 coaxial cable?

A. There is no significant difference between the two types
B. RG-58 cable has two shields
C. RG-8 cable has less loss at a given frequency
D. RG-58 cable can handle higher power levels

RG-8 has a loss of 6 dB/100 ft at 400 MHz, while RG-58 has a loss of around 12 dB/100 ft. The RG-8 has a lower loss making **Answer C** the right choice.

T9B11 Which of the following types of feed line has the lowest loss at VHF and UHF?

A. 50-ohm flexible coax
B. Multi-conductor unbalanced cable
C. Air-insulated hard line
D. 75-ohm flexible coax

This question is very general in nature, but this will give you some rough indication of the order of the loss. We just saw that RG-8 and RG-58 were in the 6 to 12 db/100 ft region, and they both are 50 Ω co-axial cables. RG-6 is a typical 75 Ω cable, and it has a loss of 7 dB/100 ft, so it is similar. Unbalanced cable will also have attenuation on the order of 10 dB/100 ft. Air insulated hard line will have a loss of approximately 0.2 dB/100 ft at 400 MHz making **Answer C** the right choice.

Chapter 10

T0 – ELECTRICAL SAFETY

10.1 Introduction

This final chapter goes into important safety concepts for proper station operation. The questions are to guide you in safely operating your equipment to protect not only you, the operator, but other people near your station equipment. This includes the transceiver and other equipment inside the shack, and people who may be near your antenna when you are operating. This *Electrical Safety* subelement has the following question groups:

A. Power circuits and hazards
B. Antenna safety
C. RF hazards

This will generate three questions on the Technician examination.

10.2 Radio Engineering Concepts

The electrical safety questions will draw upon concepts dealing with fuses, power circuits, batteries, and capacitors in power supplies.

Fuses and Circuit Breakers Electrical fuses exist to prevent a device having excessive current draw, for whatever reason, from exceeding the current carrying capacity of the wiring or the maximum input current of the device. Modern homes typically use a circuit breaker in place of a discrete fuse to protect the wiring circuits in the building. Individual devices, such as a transceiver, may have a discrete fuse in the device to protect it. Together, they will prevent fires and equipment damage. The fuse or breaker rating should not exceed the maximum safe rating for either the wiring or the equipment to prevent damage. Manufacturers typically rate house wiring for 15 A, so the fusing or the circuit breaker on the Alternating Current (AC) line should not exceed this value. Naturally, all the devices on such a circuit should not require more current than the total circuit rating.

3-Phase Circuits The commercial electrical grid in the United States (US) uses 3-phase wiring to transmit AC electricity to our homes and businesses. The wiring uses standard color codes to identify the phase wires as follows:

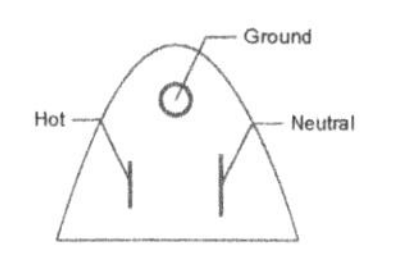

Figure 10.1: Configuration for a polarized jack for single-phase connections

Black — Phase 1 or "Hot"
Red — Phase 2
Blue — Phase 3
White or Grey — Neutral
Green — Earth ground or protective ground

We use the black, white, and green wires when we only require a single phase, such as for lighting. Our normal, polarized three-pronged plugs and jacks use the "'hot' wire on the narrow prong, the neutral wire to the wide prong, and the ground wire to the round prong. In addition to the Earth ground in the AC circuit, many circuits also include a Ground Fault Interrupter (GFI) circuit to detect when the grounding is not intact due to a problem, such as a water leak, which compromises the ground circuit path.

Batteries You are probably familiar with many types of batteries in use with modern consumer electronics. Your devices may have either rechargeable or single-use, non-rechargeable batteries. Examples of the former are Li-ion or NiCd batteries. Examples of the latter are the carbon-based batteries. In both cases, you must store and dispose of the batteries safely because the chemicals inside the battery can cause problems such as corrosive chemical leaks or fires. Another type of common battery is the lead-acid battery in your car. Many amateurs use those batteries to power their equipment for emergency communications. At high charge/discharge rates, lead-acid batteries can emit significant amounts of Hydrogen gas, which can explode if placed in a confined area.

Lightning Protection If you refer to Figure 4.1, you will see that both the shack equipment and the external antenna have specific Earth ground connections. In the shack, we tie the radio devices to a common grounding strap before connecting to the Earth ground. Both Earth grounds are 8-foot copper-clad rods sunk into the ground. This provides lightening protection for the system. The connection to the ground rod should be as short as possible and not have sharp bends or kinks in it. The wire should also have a high current carrying capacity.

Power Supply Capacitors Many power supplies contain a very large capacitor to assist in converting AC current to Direct Current (DC) current. These large capacitors also help smooth out ripples, and guard against momentary voltage drops. One problem with these capacitors is that sometimes they stay charged, like a battery, for a relatively long time after turning the equipment off as the charge gradually bleeds away. If you short the capacitor while it still holds a charge, it is the same as shorting a large battery, and you can get a severe shock or other injuries from this. You must always take care around these large capacitors.

Antenna Placement and Tower Safety Naturally, antenna placement is a consideration for operation to get the best signal performance. We also must consider antenna placement in the environment as well. The primary consideration here is that that an antenna, no matter if you mount it as a free-standing device or on an antenna support, must not contact outdoor electrical wires. The general rule of thumb is that the antenna must not come closer than 10 feet from an electrical wire. If you place the antenna on a tower, be sure that you properly extend the tower before climbing on it.

The antenna safety questions will deal with safety equipment, and safe practices for erecting antennas, especially on towers. When working on any type of outside structure, you must always consider safety practices. This includes use of ladders and tools. When working above ground, there are certain practices you should always consider such as

Buddy System — have at least one friend assist with erecting antennas and towers, so that if trouble occurs, there is someone to provide assistance or get help

Safety Equipment — even if making a quick fix, a hard hat and eye safety protection are the minimum requirements; if working on a tower, you should also include a safety harness

Tool Safety Line — consider having a safety line on each hand tool that you attach to your tool belt or harness, so that if you drop it, the tool does not hit someone on the ground below you

Radiation Hazards Radiation comes in two major classes

Ionizing Radiation — radiation capable of knocking electrons free in atoms and molecules

Non-ionizing Radiation — radiation not capable of knocking electrons free in atoms and molecules

The Radio Frequency (RF) hazard questions will deal with identifying radiation types and the maximum permissible emission limits for safe operation. RF radiation is in the non-ionizing class. While it will not cause genetic damage such as ionizing radiation can do, it can still cause burns, such as a sunburn, and one must be careful with the radiation levels. This includes not only in the shack, but for people who may come near the antenna. This is the driving reason behind the maximum permissible exposure analysis.

Maximum Permissible Exposure The government has developed a set of standards for maximum permissible exposure to RF radiation. Table 10.1 summarizes the limits. The limits are a function of operating frequency because the parts of the human body respond differently at different frequencies. The limits are based on the power density of the radiation when it encounters the human body and the *duty cycle* of the exposure. The duty cycle is the average time the radiation is "on"

Table 10.1: Maximum Exposure Limits for Occupational/Controlled Exposure (OET Bulletin 65, August 1997; see https://transition.fcc.gov/Bureaus/Engineering_Technology/Documents/bulletins/oet65/oet65b.pdf)

Frequency (MHz)	Electric Field Strength (E) (V/m)	Magnetic Field Strength (H) (A/m)	Power Density (S) (mW/cm^2)	Averaging Time $\mid E \mid^2$, $\mid H \mid^2$ or S (minutes)
0.3 – 3.0	614	1.63	100	6
3.0 – 30	$1842/f$	$4.89/f$	$900/f^2$	6
30 – 300	61.4	0.163	1.0	6
300 – 1500	—	—	$f/300$	6
1500 – 100,000	—	—	5	6

compared with the standard total observing time span of 6 minutes.

The operator should perform a radiation analysis when the operator first establishes a station, and then if the operator makes any major changes to the station, such as a change in the antenna type or placement. The Federal Communications Commission (FCC) provides a 4-page worksheet at the end of the OET65 bulletin mentioned above to assist with this evaluation. Many beginner stations will see that the worksheet shows they do not need to have the full evaluation performed because they operate at a low power level. You can find more information on the Web at sites such as http://www.arrl.org/fcc-rf-exposure-regulations-the-station-evaluation.

10.3 T0A – Power Circuits and Hazards

10.3.1 Overview

The *Power Circuits and Hazards* question group in Subelement T0 introduces you to several electrical hazards found in radio operations. The *Power Circuits and Hazards* group covers topics such as

- Hazardous voltages
- Fuses and circuit breakers
- Grounding
- Lightning protection
- Battery safety
- Electrical code compliance

There is a total of 11 questions in this group of which one will be selected for the exam.

10.3.2 Questions

T0A01 Which of the following is a safety hazard of a 12-volt storage battery?

A. Touching both terminals with the hands can cause electrical shock
B. Shorting the terminals can cause burns, fire, or an explosion
C. RF emissions from the battery
D. All of these choices are correct

As you probably suspect, 12-volt batteries do not emit RF energy, so Answer C is incorrect, which also makes Answer D incorrect. As **Answer B** indicates, shorting the terminals can cause problems, so this is the right choice. Answer A is not always true.

T0A02 What health hazard is presented by electrical current flowing through the body?

A. It may cause injury by heating tissue
B. It may disrupt the electrical functions of cells
C. It may cause involuntary muscle contractions
D. All of these choices are correct

Each of the effects mentioned in Answers A, B, and C is a possible outcome of electrical current passing through your body, so the best choice is **Answer D**.

T0A03 In the United States, what is connected to the green wire in a three-wire electrical AC plug?

A. Neutral
B. Hot
C. Equipment ground
D. The white wire

Remember: green is for ground, so **Answer C** is the right choice. The neutral and hot wires set up the 120 V reference, so they would not be good choices for where to have the ground connection. This makes Answers A, B, and D incorrect.

T0A04 What is the purpose of a fuse in an electrical circuit?

A. To prevent power supply ripple from damaging a circuit
B. To interrupt power in case of overload
C. To limit current to prevent shocks
D. All of these choices are correct

A fuse protects against electrical overloads, so **Answer B** is the right choice. The fuse cannot remove power supply ripple, so Answer A is not a good choice. The fuse does, in a sense, prevent shocks by limiting the amount of current, but this is not how we normally think of a fuse's protection property, which makes Answer C incorrect. Since Answers A and C are incorrect, Answer D is also incorrect.

T0A05 Why is it unwise to install a 20-ampere fuse in the place of a 5-ampere fuse?
A. The larger fuse would be likely to blow because it is rated for higher current
B. The power supply ripple would greatly increase
C. Excessive current could cause a fire
D. All of these choices are correct

The purpose of a fuse is to protect your equipment from drawing too much current and destroying equipment or creating a fire hazard. If the fuse blows at 5 amps, then you should not allow more current than that into the equipment to keep things safe and protected. Placing a 20-amp fuse in the line will allow too much current to flow to the equipment, and bad things can happen, so **Answer C** is the best choice. Answer A is just the opposite of the desired effect, so it is a bad choice. Answer B is incorrect engineering, so it is a distractor. Because Answers A and B are incorrect, Answer D must be incorrect as well.

T0A06 What is a good way to guard against electrical shock at your station?
A. Use three-wire cords and plugs for all AC powered equipment
B. Connect all AC powered station equipment to a common safety ground
C. Use a circuit protected by a ground-fault interrupter
D. All of these choices are correct

Each of the practices mentioned in Answers A, B, and C is good engineering practice to prevent shocks, so the best choice is **Answer D**.

T0A07 Which of these precautions should be taken when installing devices for lightning protection in a coaxial cable feed line?
A. Include a parallel bypass switch for each protector, so that it can be switched out of the circuit when running high power
B. Include a series switch in the ground line of each protector to prevent RF overload from inadvertently damaging the protector
C. Keep the ground wires from each protector separate and connected to station ground
D. Ground all of the protectors to a common plate which is in turn connected to an external ground

Answers A and B are faulty because they include switches that could allow you to remove a protection, so they are not good practice. Answer C does not make good electrical sense. Use of a common ground plate and tying it to an external ground, as in **Answer D**, represents good practice, so it is the right answer.

T0A08 What safety equipment should always be included in home-built equipment that is powered from 120V AC power circuits?

A. A fuse or circuit breaker in series with the AC hot conductor
B. An AC voltmeter across the incoming power source
C. An inductor in series with the AC power source
D. A capacitor across the AC power source

All home-built equipment should have a fuse or circuit breaker to meet good engineering practice, so **Answer A** is the right choice. The other choices are not enhancing safety, so they are incorrect.

T0A09 What should be done to all external ground rods or earth connections?

A. Waterproof them with silicone caulk or electrical tape
B. Keep them as far apart as possible
C. Bond them together with heavy wire or conductive strap
D. Tune them for resonance on the lowest frequency of operation

The Earth ground system does not need chemicals getting in the way, so waterproofing may defeat connections making Answer A a poor choice. Answer B will not help the grounding, so this is incorrect as well. Tuning for resonance will invite RF problems with your emissions, so this is not a good choice. Bonding them together with an appropriate conductor is good engineering practice, so **Answer C** is a correct choice.

T0A10 What can happen if a lead-acid storage battery is charged or discharged too quickly?

A. The battery could overheat and give off flammable gas or explode
B. The voltage can become reversed
C. The memory effect will reduce the capacity of the battery
D. All of these choices are correct

Improper battery charging can cause battery overheating and gas discharge, so **Answer A** is the right choice. Voltage reversals and memory effect are not characteristics of lead-acid batteries, so Answers B and C are incorrect. This also makes Answer D incorrect.

T0A11 What kind of hazard might exist in a power supply when it is turned off and disconnected?

A. Static electricity could damage the grounding system
B. Circulating currents inside the transformer might cause damage
C. The fuse might blow if you remove the cover
D. You might receive an electric shock from the charged stored in large capacitors

Power supplies usually have big capacitors that store a great deal of charge. If this charge is not "bled off," it can form a shock hazard, so **Answer D** is the right answer for this question. Answers A, B, and C are not electrical hazards in disconnected power supplies.

10.4 T0B – Antenna Safety

10.4.1 Overview

The *Antenna Safety* question group in Subelement T0 introduces you to safety principles necessary for radio operations. The *Antenna Safety* group covers topics such as

- Tower safety
- Erecting an antenna support
- Safely installing an antenna

There is a total of 13 questions in this group of which one will be selected for the exam.

10.4.2 Questions

T0B01 When should members of a tower work team wear a hard hat and safety glasses?

A. At all times except when climbing the tower
B. At all times except when belted firmly to the tower
C. At all times when any work is being done on the tower
D. Only when the tower exceeds 30 feet in height

Good safety practice dictates that one always wears a hard hat and safety glasses around the tower, as indicated in **Answer C**. The other choices are less safe, so they are incorrect here.

T0B02 What is a good precaution to observe before climbing an antenna tower?

A. Make sure that you wear a grounded wrist strap
B. Remove all tower grounding connections
C. Put on a carefully inspected climbing harness(fall arrester) and safety glasses
D. All of these choices are correct

Wrist grounding straps are for working on bench-level systems and not towers. Removing the grounding connections will be unsafe. With Answers A and B being incorrect, Answer D is also incorrect. Donning a harness and safety glasses, as in **Answer C**, is the best choice among those given.

T0B03 Under what circumstances is it safe to climb a tower without a helper or observer?

A. When no electrical work is being performed
B. When no mechanical work is being performed
C. When the work being done is not more than 20 feet above the ground
D. Never

The short answer to this question is "never," as in **Answer D**. The other choices are unsafe.

T0B04 Which of the following is an important safety precaution to observe when putting up an antenna tower?

A. Wear a ground strap connected to your wrist at all times
B. Insulate the base of the tower to avoid lightning strikes
C. Look for and stay clear of any overhead electrical wires
D. All of these choices are correct

Again, this is not the application for a wrist strap. Insulating will keep the grounding from working properly. Since Answers A and B are incorrect, Answer D is also incorrect. Of the choices given, avoiding overhead wires is the best choice, so **Answer C** is the correct answer.

T0B05 What is the purpose of a gin pole?

A. To temporarily replace guy wires
B. To be used in place of a safety harness
C. To lift tower sections or antennas
D. To provide a temporary ground

A "gin pole" is a pole with a pulley at the end, and one uses it to lift tower sections or antennas, as in **Answer C**.

T0B06 What is the minimum safe distance from a power line to allow when installing an antenna?

A. Half the width of your property
B. The height of the power line above ground
C. 1/2 wavelength at the operating frequency
D. Enough so that if the antenna falls unexpectedly, no part of it can come closer than 10 feet to the power wires

Each of the choices in Answers A, B, and C will not guarantee that the antenna will not contact the power line during adverse conditions like a storm. The usual rule of thumb is to make sure that no part of the antenna can come within 10 feet of a power line, so **Answer D** is the correct choice.

T0B07 Which of the following is an important safety rule to remember when using a crank-up tower?

A. This type of tower must never be painted
B. This type of tower must never be grounded
C. This type of tower must never be climbed unless it is in the fully retracted position
D. All of these choices are correct

Answer A is a silly distractor. Answer B is not good engineering practice. Answer D cannot be correct. Making sure that you retract the tower before climbing on it, as in **Answer C**, is the best choice here.

T0B08 What is considered to be a proper grounding method for a tower?

A. A single four-foot ground rod, driven into the ground no more than 12 inches from the base
B. A ferrite-core RF choke connected between the tower and ground
C. Separate eight-foot long ground rods for each tower leg, bonded to the tower and each other
D. A connection between the tower base and a cold water pipe

Ground rods need to be 8 feet long, so Answer A is incorrect. A RF choke to ground is not good engineering practice. The ground rods mean that you do not need to run a cold-water pipe to the tower. **Answer C**, with the 8-foot rods, is good engineering practice, so it is the right answer.

T0B09 Why should you avoid attaching an antenna to a utility pole?

A. The antenna will not work properly because of induced voltages
B. The utility company will charge you an extra monthly fee
C. The antenna could contact high-voltage power wires
D. All of these choices are correct

Avoiding power lines is important in safety engineering. The 10-foot rule that we just saw would be hard to meet in this case, so **Answer C** is the best choice to answer this question.

T0B10 Which of the following is true concerning grounding conductors used for lightning protection?

A. Only non-insulated wire must be used
B. Wires must be carefully routed with precise right-angle bends
C. Sharp bends must be avoided
D. Common grounds must be avoided

Sharp bends can cause problems with making good connections and they are not good engineering practice for ground wires, which makes **Answer C** the correct choice, and Answer B incorrect. Good engineering practice does not include Answers A and D.

T0B11 Which of the following establishes grounding requirements for an amateur radio tower or antenna?
 A. FCC Part 97 Rules
 B. Local electrical codes
 C. FAA tower lighting regulations
 D. UL recommended practices

Grounding requirements are set by the local electrical codes, so **Answer B** is the right choice. The others are to distract you.

T0B12 Which of the following is good practice when installing ground wires on a tower for lightning protection?
 A. Put a loop in the ground connection to prevent water damage to the ground system
 B. Make sure that all bends in the ground wires are clean, right angle bends
 C. Ensure that connections are short and direct
 D. All of these choices are correct

Ground connections are best when short and direct, as in **Answer C**. Answers A and B are not good engineering practice, so they are incorrect, which also makes Answer D incorrect.

T0B13 What is the purpose of a safety wire through a turnbuckle used to tension guy lines?
 A. Secure the guy if the turnbuckle breaks
 B. Prevent loosening of the guy line from vibration
 C. Prevent theft or vandalism
 D. Deter unauthorized climbing of the tower

In case you did not know, a turnbuckle is a relatively small mechanical device engineers use to establish tension in wires like guy wires. If the turnbuckle breaks, as in Answer A, the guy wires will become free, and not secured, making this an incorrect choice. Answers C and D are silly distractions since the turnbuckle is not a locking device nor will it keep someone off the tower. However, taut guy wires will tend to vibrate in the wind, which can lead to the turnbuckle turning and releasing tension. This makes **Answer B** the correct choice.

10.5 T0C – RF Hazards

10.5.1 Overview

The *RF Hazards* question group in Subelement T0 introduces you to basic electrical principles for avoiding RF hazards in radio systems. The *RF Hazards* group covers topics such as

- Radiation exposure
- Proximity to antennas
- Recognized safe power levels
- Exposure to others
- Radiation types
- Duty cycle

There is a total of 13 questions in this group of which one will be selected for the exam.

10.5.2 Questions

T0C01 What type of radiation are VHF and UHF radio signals?

A. Gamma radiation
B. Ionizing radiation
C. Alpha radiation
D. Non-ionizing radiation

Very High Frequency (VHF) and Ultra High Frequency (UHF) radiation is relatively low energy and is known as "non-ionizing" radiation, so **Answer D** is the right choice. Answers A, B, and C do not describe VHF and UHF RF signals, and they can cause ionization.

T0C02 Which of the following frequencies has the lowest value for Maximum Permissible Exposure limit?

A. 3.5 MHz
B. 50 MHz
C. 440 MHz
D. 1296 MHz

You derive the Maximum Permissible Exposure limit from the Power Spectral Density (PSD) for the RF signal. From the data in Table 10.1, 3.5 MHz can have a power spectral density of 73.5 $\mathrm{mW/cm^2}$, 50 MHz can have a PSD of 1 $\mathrm{mW/cm^2}$, 440 MHz can have a PSD of 1.5 $\mathrm{mW/cm^2}$, and 1296 MHz can have a PSD of 4.3 $\mathrm{mW/cm^2}$. From this, 50 MHz has the lowest permitted exposure of the choices given, and **Answer B** is the right selection.

T0C03 What is the maximum power level that an amateur radio station may use at VHF frequencies before an RF exposure evaluation is required?

A. 1500 watts PEP transmitter output
B. 1 watt forward power
C. 50 watts PEP at the antenna
D. 50 watts PEP reflected power

From OET Bulleting 65, a station operating at VHF frequencies with more than 50 W of transmitter power must perform an exposure evaluation. This corresponds to **Answer C**. Be careful with Answer D because it looks similar, but this is the power going back into your transmitter and, hopefully, more is going out of the antenna.

T0C04 What factors affect the RF exposure of people near an amateur station antenna?

A. Frequency and power level of the RF field
B. Distance from the antenna to a person
C. Radiation pattern of the antenna
D. All of these choices are correct

Each of the factors in Answers A, B, and C enter the RF exposure considerations, so the best choice to answer this question is **Answer D**.

T0C05 Why do exposure limits vary with frequency?

A. Lower frequency RF fields have more energy than higher frequency fields
B. Lower frequency RF fields do not penetrate the human body
C. Higher frequency RF fields are transient in nature
D. The human body absorbs more RF energy at some frequencies than at others

The best choice among the ones given is **Answer D** because the body absorbs the energy differently at different frequencies. The other statements are not true about RF absorption.

T0C06 Which of the following is an acceptable method to determine that your station complies with FCC RF exposure regulations?

A. By calculation based on FCC OET Bulletin 65
B. By calculation based on computer modeling
C. By measurement of field strength using calibrated equipment
D. All of these choices are correct

Each of the methods listed in Answers A, B, and C is an approved method, so the best choice to answer the question is **Answer D**.

T0C07 What could happen if a person accidentally touched your antenna while you were transmitting?

A. Touching the antenna could cause television interference
B. They might receive a painful RF burn
C. They might develop radiation poisoning
D. All of these choices are correct

This is one of the reasons why the FCC is so concerned with safety: someone could receive a RF burn, and **Answer B** is the right choice. Answer A is not really a safety concern and Answer C is technically incorrect, so they are not good choices to answer this question. Since Answer B is the only correct answer, Answer D is incorrect.

T0C08 Which of the following actions might amateur operators take to prevent exposure to RF radiation in excess of FCC-supplied limits?

A. Relocate antennas
B. Relocate the transmitter
C. Increase the duty cycle
D. All of these choices are correct

Since we are concerned with what comes from the antenna, moving the transmitter might not be effective. Increasing the duty cycle means that one is transmitting more frequently, so that will not reduce exposure. Since Answers B and C are incorrect, Answer D is also incorrect. Relocating the antenna, as in **Answer A**, is the best solution among the options given.

T0C09 How can you make sure your station stays in compliance with RF safety regulations?

A. By informing the FCC of any changes made in your station
B. By re-evaluating the station whenever an item of equipment is changed
C. By making sure your antennas have low SWR
D. All of these choices are correct

Answer A is incorrect by FCC rules, so this is not a good choice. If your station is generally in compliance, then the usual reason that your station might become out of compliance is that you have made an equipment change. This makes **Answer B** the best choice among those given here. Answer C will not change the RF emissions for safety, so it is not a not good choice to answer this question. Answer D cannot be correct.

T0C10 Why is duty cycle one of the factors used to determine safe RF radiation exposure levels?

A. It affects the average exposure of people to radiation
B. It affects the peak exposure of people to radiation
C. It takes into account the antenna feed line loss
D. It takes into account the thermal effects of the final amplifier

The duty cycle tells us the percentage of the time that the transmitter is emitting radiation. You use this in computing the average exposure, so **Answer A** contains the right reasoning. Answers B, C, and D are all listing irrelevant effects for determining safe exposure limits.

T0C11 What is the definition of duty cycle during the averaging time for RF exposure?

A. The difference between the lowest power output and the highest power output of a transmitter
B. The difference between the PEP and average power output of a transmitter
C. The percentage of time that a transmitter is transmitting
D. The percentage of time that a transmitter is not transmitting

The duty cycle is the ratio of the time spent doing the activity under consideration to the total time, written as a percentage. **Answer C** is the best match to this definition among the choices given.

T0C12 How does RF radiation differ from ionizing radiation (radioactivity)?

A. RF radiation does not have sufficient energy to cause genetic damage
B. RF radiation can only be detected with an RF dosimeter
C. RF radiation is limited in range to a few feet
D. RF radiation is perfectly safe

This is a return to an earlier topic. Non-ionizing radiation will not cause genetic damage, as in **Answer A**, so this is the right choice. The other options are not true technically, so they are incorrect.

T0C13 If the averaging time for exposure is 6 minutes, how much power density is permitted if the signal is present for 3 minutes and absent for 3 minutes rather than being present for the entire 6 minutes?

A. 3 times as much
B. 1/2 as much
C. 2 times as much
D. There is no adjustment allowed for shorter exposure times

Since you are applying the RF energy for only one-half the time during the averaging period, you can apply twice the amount of radiation and still be in compliance, so **Answer C** is the right choice.

Appendix A

Acronyms, Abbreviations, and Symbols

A.1 Acronyms and Abbreviations

AC Alternating Current

ADC Analog-to-Digital Converter

AF Audio Frequency

AFSK Audio Frequency Shift Keying

AGC Automatic Gain Control

ALC Automatic Level Control

ALE Automatic Link Enable

AM Amplitude Modulation

APRS Automatic Packet Reporting System

ARES Amateur Radio Emergency Service

ARQ Automatic Repeat reQuest

ASK Amplitude Shift Keying

ATV Amateur Television

BFO Beat Frequency Oscillator

BPF Band Pass Filter

BPSK Binary Phase Shift Keying

BJT Bipolar Junction Transistor

CCD Charge-Coupled Device

CEPT European Conference of Postal and Telecommunications Administrations

CFR Code of Federal Regulations

CMOS Complementary Metal Oxide Semiconductor

CSCE Certificate of Successful Completion of an Examination

CTCSS Continuous Tone-Coded Squelch System

CW Continuous Wave

DAC Digital-to-Analog Converter

DC Direct Current

DDS Direct Digital Synthesis

DF Direction Finding

DIP Dual In-line Package

DMR Digital Mobile Radio

DRM Digital Radio Mondial

DSB Dual Sideband

DSB-RC Dual Sideband - Residual Carrier

DSB-SC Dual Sideband - Suppressed Carrier

DSP Digital Signal Processor

DSSS Direct Sequence Spread Spectrum

DTMF Dual Tone Multifrequency

EHF Extremely High Frequency

ELF Extremely Low Frequency

EME Earth-Moon-Earth

EMF Electromotive Force

ERP Effective Radiated Power

FAA Federal Aviation Administration

FCC	Federal Communications Commission
FDM	Frequency Division Multiplexing
FEC	Forward Error Correction
FET	Field Effect Transistor
FHSS	Frequency Hopping Spread Spectrum
FIR	Finite Impulse Response
FM	Frequency Modulation
FSK	Frequency Shift Keying
GFI	Ground Fault Interrupter
GPS	Global Positioning System
HEO	High Earth Orbit
HF	High Frequency
HPBW	Half Power Beam Width
HPF	High Pass Filter
HST	Hubble Space Telescope
IARP	International Amateur Radio Permit
IARU	International Amateur Radio Union
IC	Integrated Circuit
IEEE	Institute of Electrical and Electronics Engineers
IF	Intermediate Frequency
IIR	Infinite Impulse Response
IMD	Intermodulation Distortion
IRLP	Internet Radio Linking Project
ISS	International Space Station
ITU	International Telecommunication Union
JFET	Junction Field Effect Transistor
LCD	Liquid Crystal Display

LED Light Emitting Diode

LEO Low Earth Orbit

LF Low Frequency

LPF Low Pass Filter

LSB Lower Side Band

LUF Lowest Usable Frequency

MCW Modulated Continuous Wave

MDS Minimum Discernible Signal

MF Medium Frequency

MFSK Multiple Frequency Shifty Keying

MMIC Monolithic Microwave Integrated Circuit

MOSFET Metal Oxide Semiconductor Field Effect Transistor

MPE Maximum Permissible Exposure

MUF Maximum Usable Frequency

NCS Net Control Station

NCVEC National Conference of Volunteer Examiner Coordinators

NEC National Electrical Code

NTSC National Television System Committee

NVIS Near Vertical Incidence Sky wave

OCFD Off Center Fed Dipole

OFDM Orthogonal Frequency Division Multiplexing

PCB Polychlorinated Biphenyl

PEP Peak Envelope Power

PEV Peak Envelope Voltage

PGA Programmable Gate Array

PF Power Factor

PIN Positive-Intrinsic-Negative

PLD Programmable Logic Device

PLL Phase Locked Loop

PM Phase Modulation

PSD Power Spectral Density

PSK Phase Shift Keying

PTT Push to Talk

Q Quality Factor

QPSK Quadrature Phase Shift Keying

RACES Radio Amateur Civil Emergency Service

RF Radio Frequency

RFI Radio Frequency Interference

RIT Receiver Incremental Tuning

RLC Resistor-Inductor-Capacitor

RMS Root Mean Square

ROM Read Only Memory

RST Readability-Signal Strength-Tone

RTTY Radio TeleType

SAR Specific Absorption Rate

SDR Software Defined Radio

SHF Super High Frequency

SID Sudden Ionospheric Disturbance

SNR Signal-to-Noise Ratio

SS Spread Spectrum

SSB Single Sideband

SSID Service Set Identifier

SSTV Slow-Scan Television

STA Special Temporary Authority

SWR Standing Wave Ratio

TDM Time Division Multiplexing

TLE Two Line Elements

TNC Terminal Node Controller

TTL Transistor-Transistor Logic

UHF Ultra High Frequency

UI Unnumbered Information

ULS Universal Licensing System

US United States

USB Upper Side Band

USB Universal Serial Bus

VCO Voltage Controlled Oscillator

VE Volunteer Examiner

VEC Volunteer-examiner Coordinator

VFO Variable Frequency Oscillator

VHF Very High Frequency

VIS Vertical Interval Signaling

VLF Very Low Frequency

VoIP Voice over IP

VOX Voice Operated Switch

VSB Vestigial Side Band

VTVM Vacuum Tube Volt Meter

XCVR Transceiver

A.2 Functions, Symbols, Units, and Variables

bps	Bits per second
c	Speed of Light, 299 792 458 m/s
C	Capacitance, F
dB	decibel
dB W	decibel Watt
dBm	decibel milliWatt
f	Frequency, Hz
F	Farads
G	Gain
Gbps	1 000 000 000 bps
GHz	1 000 000 000 Hz
Hz	Frequency unit
i	Current, A
kbps	1000 bps
kHz	1000 Hz
km	1000 m
m	meter
$m(t)$	Message signal
Mbps	1 000 000 bps
MHz	1 000 000 Hz
mV	0.001 V
μV	0.000 001 V
R	Resistance, Ω
R_b	Bit Rate, bps
s	second
$s(t)$	Carrier signal

V Volt

Ω Ohms

W Watt

X Reactance, Ω

Z Impedance, Ω

Made in the USA
Columbia, SC
20 July 2018